D1367535

Roots of Social Sensibility and Neural Function

Roots of Social Sensibility and Neural Function

Jay Schulkin

A Bradford Book
The MIT Press
Cambridge, Massachusetts
London, England

©2000 Massachusetts Institute of Technology

All rights reserved. No part of this book may be reproduced in any form by any electronic or mechanical means (including photocopying, recording, or information storage and retrieval) without permission in writing from the publisher.

This book was set in Palatino by Wellington Graphics.
Printed and bound in the United States of America.

Library of Congress Cataloging-in-Publication Data
Schulkin, Jay.
Roots of social sensibility and neural function / Jay Schulkin.
p. cm.
"A Bradford book".
Includes bibliographical references and index.
ISBN 0-262-19447-3 (alk. paper)
1. Cognition—Social aspects. 2. Cognition and culture. 3. Human information processing—Social aspects. 4. Psychology, Comparative. I. Title.

BF311 .S385 2001
153—dc21 00-026950

To Nicky and Danielle Schulkin, that they may always have strong bonds with kind, wise, courageous, and loyal people. And to two men who stood strong for my wife, April Oliver, during a difficult time: Hamp Oliver and Jack Smith.

Contents

Preface ix
Introduction xi

Chapter 1
Intentionality and Social Sensibility 1

Chapter 2
Cognitive and Neural Sciences 19

Chapter 3
Experiments on Social Reasoning in Primates 29

Chapter 4
Development of Social Reason in Children 59

Chapter 5
Autism 87

Chapter 6
Social Reason and Action 111

Conclusion 133

Notes 141
References 145
Name Index 195
Subject Index 203

Preface

I have always been interested in people and how we come to understand ourselves. In this book, I ask two predominant questions: How are we able to appreciate the experiences of others? And what are the cognitive and neural mechanisms that render this appreciation possible? During the last twenty years, psychologists, philosophers, and neurobiologists have made some progress toward understanding something about what enables us to know someone else's experiences.

This book cuts across several disciplines and is intended for psychologists, philosophers, and neurobiologists. In the attempt to cross intellectual boundaries, I can only hope my offense is minimal and the outreach maximal. The languages of psychology, biology, philosophy, and neuroscience do not easily cross over; some of the discussion may be insular.

For those researchers not cited I apologize. The book is not intended to be exhaustive of the literature, but fair to the essential findings. The contents of the book represent years of thought on the issue of knowledge of other minds. After being an undergraduate philosophy major with Robert Neville and Georges Rey at Purchase College, and then a graduate student in philosophy in the late 1970s at the University of Pennsylvania with Elizabeth Anscombe, Michael Friedman, and Scott Weinstein, I formally moved into the neurosciences but never abandoned my interest in philosophy. I was influenced greatly by Randy Gallistel, David Premack, Paul Rozin, and John Sabini, and a short time later I had the good fortune of teaching a course on the evolution of mind as a faculty member at Penn with John Sabini, Dorothy Cheney, and Robert Seyfarth.

This book reflects my peculiar Peirceian/pragmatist roots in philosophy and orientation to science but also builds on the work of many people in the fields of psychology, philosophy, and neurobiology. I hope that, like several other new books in this area (e.g., *Mindblindness* by Simon Baron-Cohen and *Friday's Footprint* by Leslie Brothers), this work will ignite further inquiry into the cognitive neuroscience of social sensibility. We need to discover more about the neural regions and mechanisms that may underlie learning about other people's experiences.

I want to thank my family and my friends and colleagues: Kent Berridge, Kristine Erickson, Jordan Grafman, Patrick Heelan, Jerome Kagan, François Lalonde, Peter Marler, Robert Neville, Hal Pashler, Mike Power, Jeff Rosen, John Sabini, David Weissman, and Ron Wilburn.

In particular, I want to thank Alex Martin, a cognitive neuroscientist at the National Institute of Mental Health, for our many conversations during these last eight years. He has been a great colleague and a tremendous help in this project.

Introduction

We are social animals (Aristotle), and social intelligence lies at the heart of our evolution (Jolly 1966; Humphrey 1976). Representations of objects and the ability to use tools—and to develop them while interacting with others in a social group—are the lifeline of the evolutionary ascent of humanity (Jolly 1985; Hinde 1970).

What cognitive and neural mechanisms are necessary to attain the kind of social discourse that we so easily express? One cognitive mechanism for achieving such discourse is the ability to see others in terms of their wants and desires—as intentional (e.g., Dennett 1969; Searle 1983).

The hypothesis of this book is that intentionality (the content of our beliefs and desires) is a feature in the organization of action; our wants and desires are signatures in our action (Dennett 1969). This attribution of beliefs and desires to others is also a piece of biological adaptation (Premack and Woodruff 1978; Baron-Cohen 1995) and a bodily event (Merleau-Ponty 1942; Lakoff and Johnson 1999). We explore the world, replete with cognition through bodily sensibility (Bogdan 1997; Damasio 1996). Cognition need not be an alienating event, which is a major theme in the book, particularly with regard to the understanding of other people's experiences. We evolved mechanisms to discern the beliefs and desires of others, and I do not believe that this discernment is a cultural artifact, though there may be great cultural variation in its expression (Lillard 1999).

Cognitive Adaptation

Problem solving has its roots in biology (e.g., Rozin 1976, 1998; Pinker 1997a; Tooby and Cosmides 1991, 1994). Cognitive abilities have adaptive significance (Rozin 1976; Pinker 1994), and cognitive events parse the world around us or internal to us into coherence. Of course, the world in which we live is pre-packaged in terms of meaningful events (Heelan and Schulkin 1998). We do not solely provide coherence; much of the world is delivered to us in coherent form (e.g., Gibson 1966; Carey 1985; Clark 1996/1997).

Specialized cognitive mechanisms pervade our mental architecture (Simon 1974; Chomsky 1972). There is no one all-pervasive design of the mind (though see Rumelhart 1989). Knowledge of causation, motion, natural kinds of objects (e.g., Barkow et al. 1992, 1995), or what foods to avoid or to eat reflects specific adaptations or specific adaptations that have been expanded from their initial use (Rozin 1976). Increased access to basic cognitive functions is one important mark in the evolution and ontogeny of intelligence; their devolution during pathology is the converse (Jackson 1884; Rozin 1976). As Rozin (1998) has nicely put it, "As with biological preadaptation, the basic idea of accessibility in development is that knowledge is borrowed rather then reinvented from scratch. Many of the cognitive problems that animals, including humans, have to solve are dealt with by a variety of specialized cognitive modules that are implemented by partially distinct neural circuits" (p. 117).

A philosophical conception of nature and design begins perhaps with a discussion of Aristotle, Darwin, and Dewey with regard to reason and adaptive design. Reason is demythologized and placed in the context of adaptation and evolution. A discussion of ecological difficulties and problem solving revolves around the rules that make behavioral adaptation possible—what Descartes (1628) described as the "rules for the direction of mind." Perhaps this definition should also include what Aristotle called "practical reason"—the rules for getting about in a social environment (see also Bourdieu 1990). Kant, in the *Critique of Pure Reason* (1787), asked what must be presupposed so that knowledge is possible: What categories are involved (see Kitcher 1990; Gallistel 1990)? Peirce (1878) asked how we hit upon the right hypothesis, or, rather, what the logic of hypothesis formation is (Hanson 1958).

Inquiry is an adaptation (Dewey 1916, 1925; Plotkin 1993) replete with biological-selection pressures amid real-world problems (Rozin 1976; Pinker 1997). One solution in predicting the behaviors of others is the capacity to understand others' experiences, other people's wants and desires. In other words, the intentional stance (Dennett 1978) facilitates inquiry into each other's experiences. Therefore, I emphasize in this book the preconditions for social knowledge, namely, our ability to interpret the behavior of those around us. This ability is linked to the concept of intentionality; in other words, it is the ability to attribute to others, and to see them as having, beliefs and desires in their actions, which may or may not lead to an appreciation of other people's experiences. Unlike Dennett, I take this ability to be a real feature of the information-processing system in the brain, in addition to being a useful cognitive tool in the prediction of behavior.

Pragmatism, Inquiry, Cognitivism, and Experience

The early pragmatists' recurring theme is that knowledge is tied to experience and inquiry (Peirce 1878; James 1910; Dewey 1925). The pragmatists went to great lengths to distinguish their sense of experience from that of seventeenth- and eighteenth-century empiricism (Smith 1970; Neville 1974).

Peirce (1877, 1887), the preeminent pragmatist, understood earlier than most of his colleagues the meaning of symbolic computational systems and their dependence in use on principles of interpretation.[1] His familiarity with formal logic and its history grounded his logic of inquiry (Peirce 1878b, 1889), which comprises three modes or stages: *abduction*, or the genesis of an idea; *induction* to see whether evidence supports the idea; and *deduction* of logical consequences from general principles.

Dewey (1925), influenced by Darwin (see Dewey 1910/1965), shared the view that the origin of inquiry is in the precariousness of human existence and the countless searches to recapture lost equilibrium. In the pursuit of inquiry, theories were used to guide actions, and feedback from actions served to correct theories. In Dewey's naturalistic evolutionary vision of life, the twin cardinal poles of human action were strife and resolution.

Much of what the earlier pragmatists (e.g., Peirce 1871) proposed was that cognition was endemic to problem solving of any sort, including sensory processing (see also Wittgenstein 1953). Hanson (1958), for example, defended the thesis that scientific inquiry is replete with cognition and meaning. Moreover, hypotheses are constrained (Peirce 1878a, b; Chomsky 1972); the world matters and perhaps limits our inductive capabilities (Kornblith 1985, 1993). For Hanson, following Peirce, abduction was the process of creating a hypothesis from the fallout of disputed hypotheses, within an investigatory form of life rooted in action guided by empirical consequences.

The common theme for both the pragmatist and the social constructivist is the implicit acknowledgment that knowledge is shared, in contrast to an epistemology of isolated individualism in which experience is private (Sabini and Schulkin 1994). By Darwin's (1872) or Dewey's (1925) account, for example, emotional expression evolved not in order to display one's insides to others but as a means of carrying out certain pragmatic, social actions. On that account, emotional expressions are tied to communicative gestures in social contexts, rather than to isolated internal states (Sabini and Silver 1982; Fridlund 1991). Social practice and knowledge are fundamental to any theory of mind (e.g., Wittgenstein 1953; Geertz 1973; Mead 1934).

The Contents of the Book

In our species, social reason underlies the ability to get along or not, to deceive and manipulate others, to understand the beliefs and desires of others in complicated social and evolving environments. One feature in the socialization process of reason is the discernment of the beliefs and desires of others. This process limits social or psychological isolation (chapter 1).

Social reason is about others. In the assessment of a social event, the question is often: Who is out there, or what are other people are about? Answering this question is a daunting task, yet we do it easily to varying degrees, unless we are impaired or arrested in development (e.g., autism; Baron-Cohen 1995).

For humans, the ability to attribute beliefs and desires and to predict correctly what others will do is one cognitive adaptation (Dennett 1987, 1996). However, it also creates a vulnerability perhaps to overattribute mental content to inappropriate objects (e.g., the sun, animism). But along with the overthrow of narrow behaviorism and the legitimization of mentation, there also needs to be a recovery of trying to capture the experience of others. Theorizing about others means getting a sense about their experiences (chapter 2).

Social reason is not, however, synonymous with the ability to attribute intentions to others (e.g., Aronson 1972, 1995; Sabini 1992; Shweder 1991). There obviously exists a rich social and cognitive life without this ability. Along with syntax, social reason is a major feature that underlies our evolution.

The animal mind interprets the world relative to a background framework (chapter 3). There is nothing mythical about this ability. The question for cognitive ethologists is: What sorts of categories are presupposed for animal adaptation (Gallistel 1990; Hauser 1996; Shettleworth 1999)? The range of interpretation reflects their evolutionary history. When one construes interpretation beyond a narrow conscious and human linguistic context, then the notion of interpretation is basic to animal minds. Animals are trying to generate coherent worlds in which to function, in which to gather meaning in order to survive or at least to reproduce, avoid predators, find food sources, and so on (Smith 1977; Marler 1961; Marler, Evans, and Hauser 1992). Sometimes there need not be much interpreting because the world is handed to them prepackaged with meaning (Gibson 1982; Clark 1996/1997; Heelan and Schulkin 1998). Andy Clark

(1996/1997) has nicely described this common event as the mind-body and world contributing as "equal partners" (45). The world is already quite ordered, but we nevertheless need a background framework in which to code the coherence (see also Bogdan 1997; Shettleworth 1999).

I review the literature in nonhuman primates (chapter 3) regarding the extent to which intentional understanding of the world is to be found in several species (Cheney and Seyfarth 1990b; Hauser 1996). There is some evidence in laboratory experiments that there are psychological degrees of intentional systems (Premack and Woodruff 1978; Gomez 1996; Tomasello and Call 1997; Byrne 1995; chapter 3), though by no means is it incontrovertible in laboratory experiments with chimpanzees (Povinelli and Eddy 1996a; Heyes 1998).

A growing literature in the last ten years asserts that theories of mind—that is, the ability to attribute beliefs and desires to others—emerge early in childhood (e.g., Astington, Harris, and Olson 1988; Wellman 1990; Gopnik 1993; chapter 4). An aberration of this ability to "see" other people with beliefs and desires is found in at least one known pathological condition—autism (Leslie and Frith 1990; chapter 5). The focus in this literature is on our ability to recognize others and to begin to adapt to a wide array of beliefs and desires that reflect different social environments.

I present the findings of numerous investigators who suggest that the obstacles autistic people face in terms of social reason stem from impairments in interpreting other people's beliefs and desires (Baron-Cohen 1995; Leslie 1987). What autistic individuals may lack is an ability to understand other people's experiences. This hypothesis, however, is not without limitations because autistic individuals have impairments other than that of social knowledge (e.g., Perner 1993, 1995; Frith, Morton, and Leslie 1991; Tager-Flusberg, Baron-Cohen, and Cohen 1993/1999). Nonetheless, the animal literature, when taken with the developmental literature (Gopnik and Meltzoff 1997; Flavell 1999), suggests that there may be a neural system linked to parsing our social space in terms of intentional systems (Brothers

1990; Perrett and Emery 1994; Adolphs 1999b). In fact, we know something about regions of the brain that are active when we come to understand something about the beliefs and desires of others (Baron-Cohen et al. 1999; Frith and Frith, 1999).

Regions of the brain that include the prefrontal cortex (in addition to other cortical sites, such as the temporal lobe), amygdala, and the basal ganglia may underlie the structure of intentional action (e.g., Brown and Marsden 1998; Graybiel 1997; Damasio 1994; Baron-Cohen et al. 1999; chapters 1, 3, 5, 6). These same regions may be active when discerning others as intentional actors. Action and mentation evolved together. Perhaps it is not counterintuitive that many of the same regions of the brain may be active when one performs an action and when one sees someone else do it. One anatomical representation, from the Rhesus macaque (*Macaca mulatta*) brain, depicts regions of the brain, noted above, that are critical for intentional action, the imagination of intentional action, and perhaps the recognition of intentions in others (Rolls 1999; Perrett and Emery 1994; Brothers 1994). (See figure I.1.)

I also suggest (chapter 6) that the organization of social behavior or action is replete with cognitive architecture, as are emotion systems (e.g., Parrott and Schulkin 1993a; Sabini 1992; Lane et al. 1999). In addition to the social structure that anchors us (e.g., Clark 1996/1997; Heelan and Schulkin 1998), unconscious cognitive structure pervades the organization of action (e.g., Lakoff and Johnson 1999; Gallistel 1990). Perception and sensorimotor (bodily) responses are organized by regions of the brain that are essentially information-processing systems (e.g., Lashley 1958; Jeannerod and Decety 1995; Hari et al. 1998; Martin 1999). In other words, information-processing systems in the brain pervade the traditional characterization of the sensory—the body or action; cognition pervades every level of the neural axis. There is no mind/body split on this view, no Cartesian mythology to be overcome, which Damasio (1994) has suggested has plagued the field. Moreover, results from cognitive neuroscience suggests the possibility that both intentional action and the perception of intentional action recruit many of the same

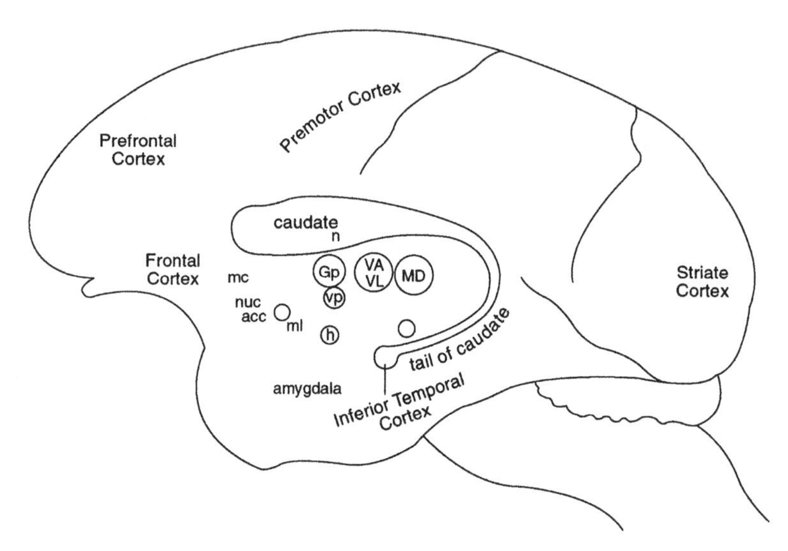

Figure I.1
A lateral view of the brain of the macaque monkey. Abbreviations: Gp, globus pallidus; h, hypothalamus; vp, ventral pallidum; MD, mediodorsal nucleus of the thalamus; VA/VL, ventrolateral thalamic nuclei nc, nucleus accumbens (adapted from Rolls 1999).

underlying information-processing neural systems (e.g., Rolls and Treves 1998; Brothers 1997; Baron-Cohen et al. 1999; Rizzolatti and Arbib 1998).

In the conclusion of the book, I return to the issue of social reason and our evolution. Social reasoning in the last five million years underlies our evolution. Social reason is about what others do—understanding their actions and predicting their behavior. Social reason underlies our own ability to adapt to the world in which we find ourselves.

Chapter 1
Intentionality and Social Sensibility

Intentional attributions, linked to the prediction of behavior and the interpretation of the experiences of others, may have been a fundamental adaptation in the evolution of social cognition. We understand others, in part, by their beliefs and desires. There is nothing mythical about this view, which has floated in many intellectual quarters. Therefore, the hypothesis is that one important feature in social reasoning is parsing the world into intentional agents: What are people's desires and beliefs? What do they want? Intentional attribution encourages meaning in social relationships (Jaspers 1913/1997). It encourages connections to others, but also deception and manipulation.

Meaning is in part housed in the world around us (e.g., Schutz 1967; Wittgenstein 1953), and we adaptively scaffold to the inherent structure of the worlds to which we are adapting and in which we are living (e.g., Gibson 1979, 1982; Clark 1996/1997). Our everyday life world (e.g., Schutz and Luckmann 1973; Mead 1936; Johnson 1987)—so rich in meaning and grounded in practice—is replete with human intentionality and is represented in a rich set of cognitive structures in the brain (e.g., Baron-Cohen et al. 1999; Brothers 1997; Adolphs 1999).

In what follows, I provide a brief historical background on the concept of intentionality, followed by a discussion of social meaning and then finally of how the brain might discern social meaning. In my view, intentionality is a "real" information-processing system in the brain built perhaps from the perception of the direction of action patterns. It has evolved into a fundamental way in which we learn about the experiences of others. It is important first to get a sense of the concept of intentionality, then an understanding of how meaning is in the world that we inhabit and to which we adapt and how the brain may function toward objects and behavioral expression in discerning meaningful events.

Historical Roots of the Concept of Intentionality

A medieval distinction between *esse naturale* and *esse intentionale* underlies this discussion. The first has to do with things or kinds, the second with intentions or thoughts, or with objects in thought (see e.g., Lyons 1995).

Intention was construed by Franz Brentano at the turn of the twentieth century as one of the essential features of mind. He described intentionality in terms of perception looking inward (1875, p. 34). Husserl (1913/1931) elaborated on Brentano, linking intentionality to the structure of consciousness. But the context of phenomenological description was abstract and in the end divorced from experience. Perhaps the more plausible among Husserl's exploits in phenomenology is his claim that all evidence for theory is tied to intention (see also Sellars 1963, 1968).

It was clear then that intentionality is tied to representational capacity; propositional attitudes became the academic parlance of discussion (Russell 1950). But, to be sure, something can be representational and not be intentional. I do not think the converse holds, however. A computational/representational theory of mind is an essential part of the cognitive revolution and figures importantly in any realistic theory of intentionality (Rey 1999).

There are two senses of the concept of intentionality: *intension,* which is the sense and reference or extension of a proposition, and *intention,* which is "aboutness."

Brentano used the concept of intentionality with a *t* or *s* to reinforce traditional Cartesian dualism. Of course, what he was right about is the pervasive characteristic of intentionality in human experience.

An important distinction fundamental to a theory of meaning (Frege 1892) is between intension and extension. The extension, the object, can be the same and obviously have a number of intended meanings. Moreover, depending on how one understands the terms, intensions are preliminary to extensions (Sellars 1968). As Sellars so nicely noted, "extensions are limiting cases of intentions and cannot be understood apart from them" (p. 77).

Intentions are shared, part of the fabric of the community, despite the fact that they are embodied in individuals. Anscombe's (1957) book *Intention,* following Wittgenstein (1953), was mostly concerned with the right use of the term. There was a public and social language in which intention and its attribution were legitimate; the language was tied to rational human action (e.g., Aristotle). The context was social, with an emphasis on being careful about the use of language (Ryle 1949; Austin 1970). The emphasis was also on intersubjective discourse—the practices and habits that ground us in everyday life in attributing intentional action to someone. But Rylian emphasis always bordered on behaviorism.

What precipitated this stance about the primacy of intentional discourse was the endless confusion about looking inward when there was nothing there to look in on—the flaw of both classical rationalists or empiricists. According to the Ryle/Anscombe view, intentionality was a marker among sentences (Dennett 1969). Intentionality in either sense was essentially linguistic or at least fundamental to the linguistic behavior of other individuals. As Dennett (1969) stated his position early on, "intentionality is not a mark that divides phenomenon from phenomena, but sentences from sentences" (p. 27).

Linguistic behaviorism (Anscombe 1957; Ryle 1949) and the attribution of intentional discourse to predict behavior became a dominant position; In other words, intentional attributions are useful; they provide one with predictive prowess. For example, in "interactions" with a computer, how does the attribution of intentionality play a role in what one thinks a rational agent will do next (Dennett 1978)?

But it is one thing to assert, *rightfully,* that language users understand much about the concept of intentionality; it is quite another to say that only our species is intentional, that only language users are intentional, and that, therefore, prelinguistic infants are not intentional. The latter position is not supported by some of the data on neonates (e.g., Meltzoff 1995) or by those of us who have paid attention to our children.

Perhaps the ability to perceive others as intentional is both a biological/cognitive adaptation and a core concept in our sense of ourselves and in our representational abilities. It may be a piece of mental adaptation (Pinker 1997) that is fundamental to a theory of human action (Lakoff and Johnson 1999). In addition, the concept of intentionality functions as a theoretical concept in our folklore or our folk symbolic sense of ourselves (see also Lyons 1995). Intentional discourse is a predictive tool and functions in the language games of our communities. The concept is part of our everyday folk psychology (Bogdan 1997). Accounts of intentional agents are one important way in which we account for others' *experiences.*

Consider the range of intentional judgments in playing basketball. The genius of Michael Jordan is both his great athletic prowess and his ability to predict where his teammates will be. His ability to "see" what is going on traverses the playing field. The sensorimotor responses are inherently cognitive, intentional, anticipatory, and responsive to others. When played with great excellence, the game is social and cooperative for those playing on the same side and deceptive to those on the other. It is inherently replete with intentionality both in behavior and in the attribution to others.

Intentional Systems

Intentional attribution in everyday social life perhaps reflects an underlying communicative social competence (Habermas 1990; Tirassa 1999); Competence is distinguished from performance (Chomsky 1972) in distinguishing underlying theory and normativeness from actual behavior.

The mechanisms for being an intentional agent or for attributing intentions to others are not necessarily personal—that is, conscious; either the attribution to others or the mechanisms that underlie the intentional stance are subpersonal (Dennett 1969, 1978; see also Boden 1970). The intentional stance is part of the hardware of reason and a forceful tool in the prediction of behavior. We want to understand the design of a system and its mechanisms.

Intentional systems have several layers. First-order intentional systems refer to events, but not necessarily to beliefs and desires; second-order intentional systems are about beliefs and desires; and third-order intentional systems are about wanting others to believe what one wanted. A question for the ethologist or the developmental psychologist, as Dennett and others phrased it, is: To which species and at what age do these different sorts of events occur (chapters 2 and 3)? What significance does intentionality play in the organization of action and in successful problem solving?

The distinction that Searle (1983) made between intrinsic and nonextrinsic states of intentionality is important if one holds onto the idea that intentionality is a real property of the person and not just a good tool, which may or may not be a property of the person. Both concern directness, representation. Both are about the fact that we might have different experiences looking at the same physical objects; recall the famous example of the duck (originally Jastrow, the duck-rabbit; see Wittgenstein 1953, figure 1.1, and Brown 1977). Searle's distinction also highlights the role of theory (the use of a framework for seeing embodied in the brain) in perception. All "seeing" reflects a background body of knowledge and expectations (Hanson 1958), and from a biological point, our perceptual systems reflect the environment to which we adapted (Shepard 1992).

Searle (1983) notes that (1) intentions are secreted by the brain analogous to insulin being secreted by the pancreas, and that (2) "intentional states do not function autonomously" (p. 190). Or, as he stated a few pages later in one of his books, "intentionality occurs in a coordinated flow of action and perception" (1992, p. 195). He then states that intentional discourse occurs amid a background, which I would suggest in part consists of the social practices, the external order that is all-pervasive (Clark 1996/

Figure 1.1
Picture of a possible duck perception (Wittgenstein 1953/1958).

1997), and the frames of references (biological/cognitive/social) that we presuppose.[2]

Intentional explanations are functional, information-processing systems (Dretske 1981, 1995) whose psychological necessity resides within a biological evolutionary context (Millikan 1984, 1998). They facilitate social understanding, prediction, and deception. Intentionality functions within interpretive contexts (Heelan and Schulkin 1998) with "meaning" inherently social amid our participation in the community.

Intention and Meaning: The Social World

Social cognition (Fiske and Taylor 1991) underlies the ability to accommodate or assimilate to the customs of our culture (Piaget 1952, 1971). Meanings are shared, and social intentional relationships are essential to communicative discourse (Grice 1957). What it was that you intended by *x* is as basic as any part of our vocabulary. Our ability to understand each other's experiences and to manipulate one another or not, as well as the degree to which we do or do not, are tied to the intentional stance (Dennett 1987). Our individual experiences are importantly sustained, as I indicated in the introduction to the book, by the meanings that we derive together and are connected to one another in everyday life (e.g., Jaspers 1913)—the stuff that binds us and renders our life meaningful.

Meanings are embodied in the practices of everyday life (Goffman 1959, 1971) and in our linguistic utterances (Austin 1970; Searle 1983). Meaning is determined in the interaction of mind adapting to environments. Sometimes it has to do with minds discerning other minds and predicting and understanding events as a function of this discernment. Meaning is thus neither in the head nor strictly in the environment; it is in the interaction between the two (Dewey 1925; Lakoff and Johnson 1999). But they are real. As Peirce (1871) stated it, "This realistic theory is thus a highly practical and common-sense position" (p. 488).

Social knowledge often requires interpreting the intentional meaning of others; often what is at stake is what one thinks the other intends to do (Grice 1957; Schiffer 1972). What is at stake in social discourse is capturing the meaning of others' utterances and their actions, which requires the attribution of beliefs and desires. Meaning pervades this activity. There is nothing mystical about it; there is, however, a question whether one can imagine that the brain computes this sort of thing without recourse to folk psychology and the theory of speakers' intentional meaning that pervades this psychology (cf. Stich 1983; Churchland and Churchland 1998).

Intentional attributions are linked to social knowledge or understanding (Barresi and Moore 1996). They force social parsing of other people's wants and beliefs. Moreover, intentions are a basic attribution that we perform and see in others. Intention can be linked both to ongoing action or to the anticipation of action (Bratman 1987; Searle 1983). Various degrees of instantiating intentional stances have been described; the important point from the social perspective is taking other minds into account, perhaps in the context of goodwill, deception, or manipulation, but ultimately in the context of trying to discern and predict behavioral outcomes and thus guide one's behavior. Nonetheless, one can have a belief about an event or object regardless of what anybody does or thinks—even if one were the last person on Earth.

Brain Function and Meaning

Both the cognitive revolution and the emphasis on biological adaptation have fostered the view that there is a diverse array of cognitive mechanisms in learning (Rozin 1976) and in memory

(Squire 1987; Tulving and Markowitsch 1998). Semantic processing reflects different cognitive and neural mechanisms (e.g., Squire and Zola 1996). More than one system appears to underlie semantics in the nervous system (e.g., Warrington and Shallice 1984; Warrington and McCarthy 1987). Moreover, from an anatomical point of view, representations of object knowledge, like most functions in the brain, are not simply localized in one part of the brain, but are distributed across the neural axis (e.g., Small et al. 1995; Spitzer 1999; Rolls, Treves, and Tovee 1997; Wise et al. 1991; Ishai et al. 1999). But the representations are not random; they reflect the underlying organization of the nervous system as well as its evolution and function. One important insight into the brain is that semantic processing reflects regions of the brain linked to action and perception (Martin 1998; Jeannerod 1999; Rizzolatti and Arbib 1998).

Consider briefly some illustrative basic facts about neural structure and function with regard to object and social perception. Experiments by Roger Shepard and his colleagues (Shepard and Cooper 1982; Shepard and Metzler 1971) demonstrated that whether a visual rotation is imagined or an object is actually looked at, the time period needed to do so reflects the size of the object. Moreover, we now know that similar neural circuits are also activated when the object is imagined or viewed (Farah 1984; Kosslyn 1994). Imagining is the process of creating brain stimulation internally similar to what would be created from external stimulation. In other words, the neural structures active in imagining objects appear similar to those structures active when looking at them (figure 1.2; see Kosslyn 1980, 1994).

For example, using positron emission topography (PET) to measure regional blood flow, it has been observed that the brain regions that are active when one looks out on a visual stimulus are also active when one imagines the object (Kosslyn 1994). These regions include the left angular gyrus, dorsal lateral prefrontal cortex, right superior parietal lobe, area 17 of the left hemisphere, fusiform gyrus in both hemispheres, and the right inferior parietal lobe (Kosslyn, Alpert, and Thompson 1993). These regions of the brain play differential roles in determining

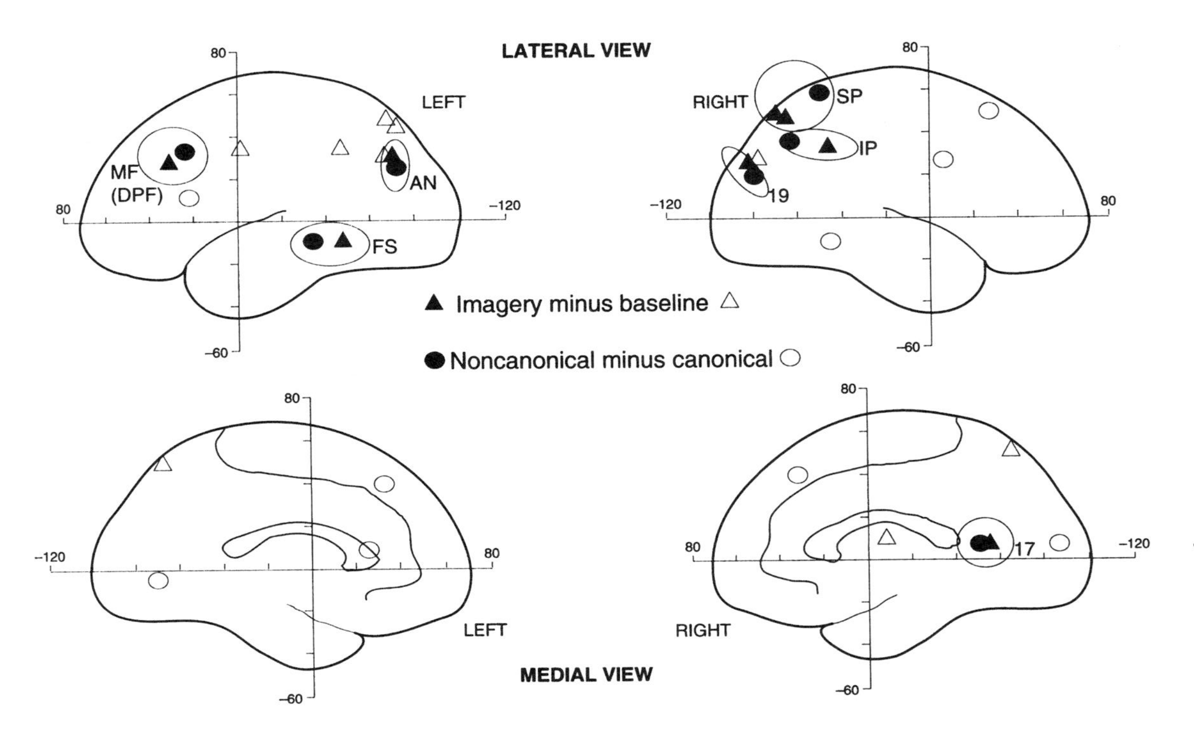

Figure 1.2
The circles indicate foci of activity that were evident when blood flow in the canonical pictures condition was subtracted from blood flow in the noncannonical picture condition; the triangles indicate foci of activity that were evident when blood flow in the baseline condition was subtracted from blood flow in the imagery condition; the black symbols indicate areas activated by both types of processing. Abbreviations: MF, middle frontal area; AN, angular gyrus; FS, the fusiform gyrus; SP, superior parietal cortex; IP, inferior parietal cortex; 19, area 19; 17, area 17 (after Kosslyn, 1994; Kosslyn, Alpert, and Thompson 1993).

visual objects and their meaning, and also in imagining those objects (see also, e.g., Roland and Gulyas 1994; Cohen et al. 1996; Zeki, Watson, and Frackowiak 1993; Moscovitch, Behmann, and Winocur 1994).

Consider another example in the auditory system. In a functional magnetic resonance imaging study (fMRI), subjects were presented with auditory words through headphones, and in another experiment the same individuals were asked to identify the words with silent lipreading (Calvert et al. 1997). The temporal cortex was activated during both the heard speech and the lipreading. Other regions of the brain linked to visual motion were activated (i.e., extrastriate cortex and inferoposterior temporal lobe [Calvert et al. 1997]). (See figure 1.3.)

In studies using PET to measure blood flow or neural activation, subjects were asked to imagine grasping objects (e.g., Decety, Perani, and Jeannerod 1994; Stephan, Fink, and Passingham 1995; Parsons et al. 1994). Significant activation of regions of the brain concerned with movement was apparent. For example, Brodmann area 6 in the inferior part of the frontal gyrus of both cortical hemispheres was active when subjects were asked to imagine grasping an object. The anterior cingulate and the parietal cortex were also activated. In addition, both the caudate nucleus of the basal ganglia and the cerebellum were activated. In further studies using neuromagnetic methods to measure cortical activity, the primary motor cortex was active both when subjects observed simple movements and when the subjects performed them (Hari et al. 1998). Motor imagery is replete with cognitive structure and is reflected in the activation of

Figure 1.3
Gray rectangular areas indicate brain areas activated by silent lipreading in (*A*) and its replication (*B*) overlaid on areas activated during auditory speech perception in experiment 1 (black areas). White areas indicate regions activated in common by silent lipreading and heard speech. These generic brain-activation maps are superimposed on spoiled GRASS MR images centered at 1mm (left), 6.5mm (center), and 12mm (right) above the intercommissural line. The left side of each image corresponds to the right side of the brain (Calvert et al. 1997).

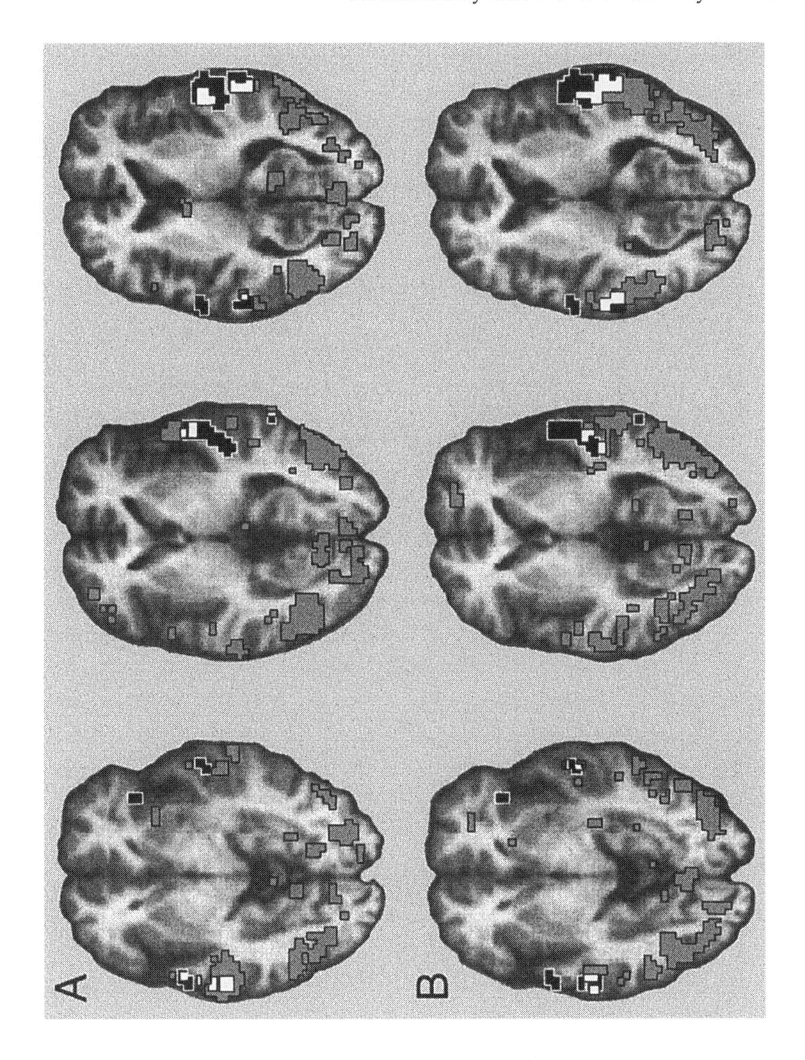
A
B

neural circuitry (see also Rizzolatti and Arbib 1998; Decety 1996; chapter 6). (See figure 1.4.)

My colleagues at the National Institute of Mental Health, Alex Martin and his coworkers, have been studying which regions of the brain are active in identifying specific classes of objects (Martin et al. 1995, 1996; see also, e.g., Perani et al. 1995, 1999; Tranel et al. 1997; Bookheimer et al. 1995). Their hypothesis is that the semantic encoding will reflect the activation in regions of the brain that are known for processing specific features of objects (Martin et al. 1995, 1996). Specific regions of the brain were activated when subjects were asked about their knowledge of color words versus their knowledge of action words. In studies using PET, action words generated greater activity in the left middle temporal gyrus (Martin et al. 1995). This region of the brain, interestingly, is just anterior to a region linked to the perception of motion (see, e.g., Rolls 1999; Perrett and Emery 1994). (See figure 1.5.)

Using PET to measure regional cerebral blood flow, the researchers asked subjects to identify drawings of animals and tools. These classes were chosen because the distinctions we make among four-legged animals usually rest on their physical differences (size, color, etc.), whereas our distinctions among tools are based on their functions (Martin 1999). Naming animals (but not tools) activated the medial aspect of the occipital lobe. Perhaps this result reflected the fact that the task made demands on early visual processing. The activation of the occipital cortex reflects a reactivating of primary visual areas, which may arise from the need to identify an object using the relatively subtle distinctions of physical features. (See figure 1.6.)

Naming tools (but not animals) activated the left middle temporal gyrus, the same region that is activated in generating action words associated with objects (Martin et al. 1996). The authors suggest that this area may organize stored knowledge of visual motion and their use.[3]

The authors conclude that identifying tools may be partly mediated by the areas of the brain that mediate knowledge of object motion and use, and are close to sites that are active when

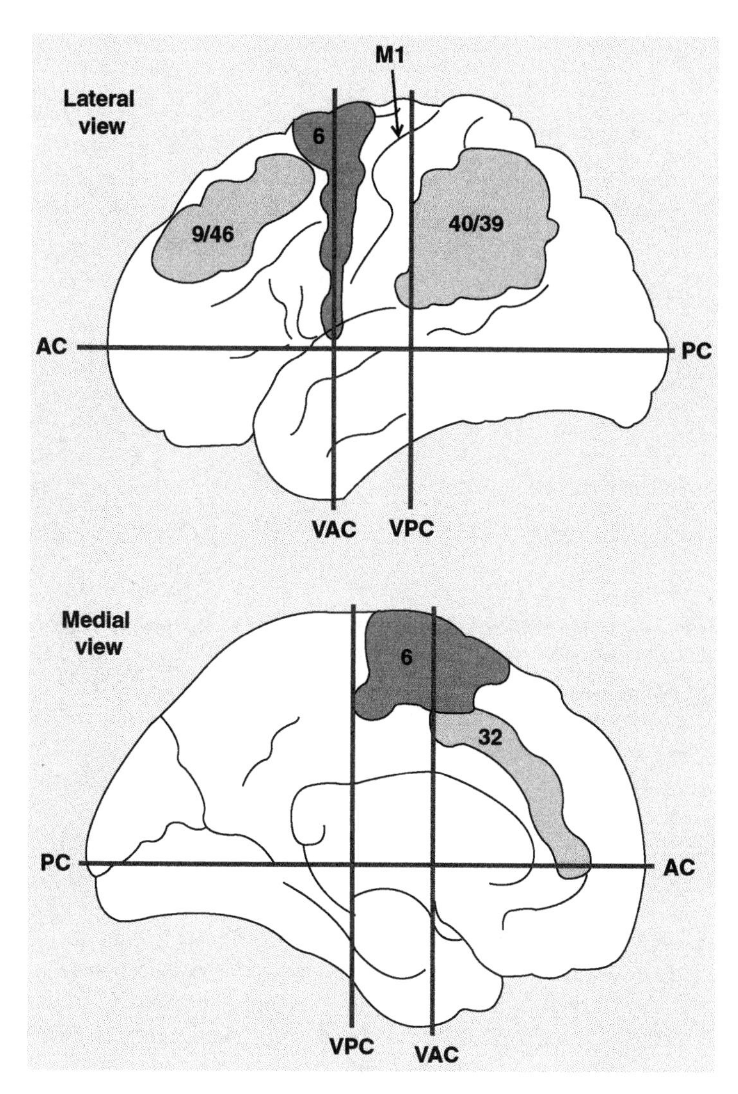

Figure 1.4
Pattern of cortical activation during mental motor imagery. The main Brodmann areas activated during motor imagery have been outlined on schematic views of the left hemisphere. Note the involvement of the premotor area 6, without involvement of primary motor cortex (M1). The AC-PC (anterior to posterior commissure) line defines the horizontal reference plane in the magnetic resonance imaging scan The vertical line passing through the AC (VAC) defines a verticofrontal place. VPC is the vertical place passing through the PC (after Jeannerod and Decety 1995).

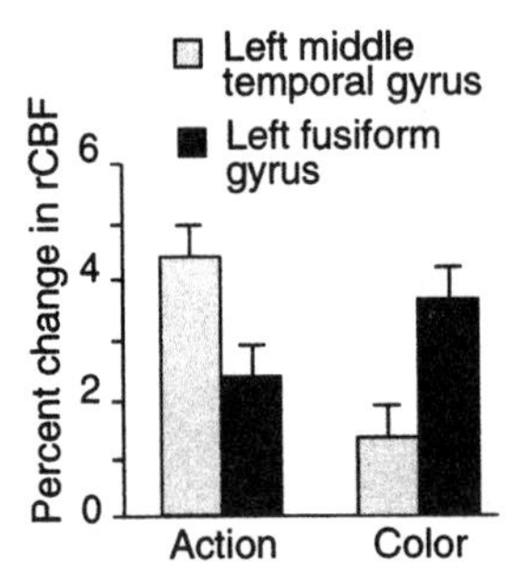

Figure 1.5
Percent change in rCBF, relative to object naming, at the site of peak activity in the left middle temporal gyrus (open bar) (–50, –50, 0) and left fusiform gyrus (closed bar (–46, –46, –12). Bars represent mean percent change in rCBF ± SEM (after Martin et al. 1995).

we are perceiving motion and using objects. In this context, perception and action are joined (e.g., Martin et al. 1995, 1996; Chao et al. 1999; see also Lakoff and Johnson 1999). Again, that does not mean they are not separable (Jeannerod and Decety 1995; Rizzolatti and Arbib 1998). Obviously, they are and importantly so, but these sorts of anatomical findings point to common underlying neural mechanisms.

The important evolutionary and functional point about the brain is that viewing the features of an object activate both visual and motor regions within the brain (chapter 6). One region reflects sight, the other action or movement. Both are part of the semantic neural mechanisms. Both can be active whether we are looking at movements or moving ourselves through space. There appears to be a predilection to discern motion; action categories (e.g., animate or inanimate; Premack 1990) and functional relationships (use of an object) may figure importantly in discerning motion and causation (e.g., Ristau 1998; Hauser and Carey 1998).

An emphasis is on the body's appropriation of objects and their use (Dewey 1925; Merleau-Ponty 1942; Lakoff and Johnson 1999). The body is a vehicle of knowledge, replete with

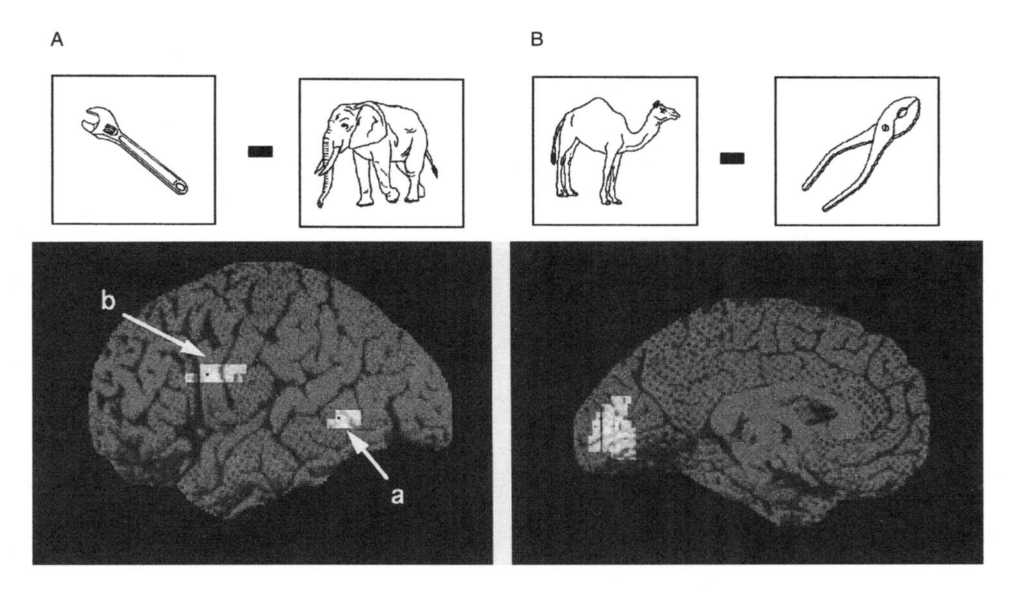

Figure 1.6
(*A*) View of the left side of the brain showing areas in the left posterior temporal lobe (a) and left premotor cortex (b) that were more active when subjects silently named pictures of tools than when they silently named pictures of animals. (*B*) View of the inner (medial) surface of the left side of the brain (right side removed) showing the region of the occipital lobe that was more active when subjects silently named pictures of animals than when they silently named pictures of tools (Martin et al. 1996, Martin 1998).

cognitive structure for knowing what is around and what to attend to, learn from, and respond to. This sense of body knowledge is well represented in the brain and is part of the organization of intelligent action (Damasio 1994). The sensorimotor organization is also replete with representations that are cognitive (Jeannerod 1997, 1999; Rizolatti and Arbib 1998). There are no bare bodily events. The class of representations is larger than simply propositional ones. Cognitive structure is pervasive.

The anatomical/functional relations have relevance to considerations about regions of the brain and social knowledge, as we will see in later chapters. It may be the case that my acting in an intentional way, or imagining that I am, and my attributing intentional action to you activate the same brain regions, and that these brain regions may contribute to the organization of action (e.g., Gallese and Goldman 1998). The neural and cognitive processing is performed outside of our awareness. We do not have privileged access to the mechanisms that underlie semantic processing in the brain.

Unloading Meaning from the Mind/Brain Back to the World

Intentional action is often taken in the context of others, potential others, and imagined others (deceiving others, helping others). Even the most reclusive of us imagines others or lives with others, at least to the extent in which perhaps we are still intentional and have a background of socialized ambiance—rejected or not. Moreover, the worlds we inhabit and adapt to already contain well-worked practices that pervade and scaffold us into a world we embody (e.g., Peirce 1889; Clark 1996/1997).

To relieve the brain from having to store so much information, we have two choices available to us. First is the Gibsonian route: meaning comes in the information that we directly receive from the world in which we live. We derive meaning through our senses; we cannot attribute meaning without the sensory input, the external information. But Gibson (1966, 1979) emphasized action patterns that are released by invariant properties of the world in which we evolved. He was short on empirical data, however. Moreover, the mind is not barren; the senses themselves are within cognitive systems. Gibson unloaded too much of the mind, leaving nothing. He was referring only to simple psychological mechanisms (e.g., Schacter 1992; Schacter and Cooper 1993), simple perceptual/structural abilities.

Moreover, some categorical distinctions that we express are more easily learned than others. We are prepared to learn some kinds of events more easily than others (Garcia, Hankins, and Rusinak 1974; Rozin 1976). Perhaps disinctions such as intentional/nonintentional, causal/noncausal, and animate/inanimate objects and their meanings reflect this preference (e.g., Ristau 1998; Hauser and Carey 1998; see chapters 3 and 4).

To simplify our cognitive abilities, we learned to tap into the meanings and stability of natural kinds (see Carey 1985; Kornblith 1993). Therefore, there need not always be a lot of information processing. In this sense, meaning comes in the reception itself (e.g., Gibson 1979; Clark 1996, 1997). Gibson is vague, but his point from the biological side is that the world is already coherent; it has inherent regularity or clusters of regular patterns that provide coherence for which the brain is prepared to receive information. The same holds for social kinds. Stable entities are pervasive, and the meaning is out there in these stable entities. There is less need for processing (Clark 1996/1997; Heelan and Schulkin 1998).

Summary

Intentionality is indeed one mark of the mental. Intentionality is knotted to our representational capacity; information is exchanged in the communities in which we reside and participate. Inquiry geared toward settled routines is embodied in pervasive social practices. The events are intersubjective, and the social practices are embodied in the communities in which we reside and participate. The knowledge is coded in coherent clusters of meaning; the meaning is prepackaged for use.

Moreover, we did not evolve in isolated vacuums, and we may be more readily prepared to code some kinds of objects and screen others (see chapter 3; Carey 1985; Keil 1992). A commonsense realism pervades our natural inclinations about the reality of beliefs and desires of others (Ristau 1998), which, although containing varying degrees of opaqueness, are nonetheless a predictive tool of their behavior. Moreover, there is more than one semantic system, analogous to the fact that more than one kind of learning system is organized in the brain (Rozin 1976; Shettleworth 1971; Seligman 1970).

Neural circuitry in the brain underlies semantic processing that reflects features of the object perceived, imagined, and perhaps acted upon or used. Neural circuits are often task or function oriented. Finally, imagining an intentional action (to be developed in subsequent chapters) and actually viewing someone may recruit similar neural regions, analogous to other systems in the nervous system.

Chapter 2
Cognitive and Neural Sciences

As I indicated in the preceding chapter, understanding the intentions of others is not strictly an epistemological dictum. Under a number of conditions, we want not only to predict behavior as accurately as possible, but also to understand something about the experiences of others. This is a fundamental feature of social knowledge, and our biological machinery is prepared early on in ontogeny to perform this function. One result from learning from other people's experiences is an increase in the ability to predict and influence behavior. Moreover, understanding the experiences of others is a fundamental way in which we learn about the world or get a foothold in our social milieu. Thus, one result of this ability to parse the world in terms of beliefs and desires is that it provides us with a window into the experiences of others (Baron-Cohen 1995). There is nothing foolproof about it. People inhabit understanding of themselves and others with both vibrant clarity and utter opacity. (See figure 2.1.)

The cognitive revolution (e.g., Gardner 1985; Gazzaniga 1995) needs to preserve the emphasis on experience, perhaps originally formulated by the early pragmatists (e.g., James 1912; Dewey 1925; Mead 1934). These early pragmatists viewed psychological functions in terms of their adaptive value, and they saw an appreciation of others' experiences as a vehicle for learning about the world, for engaging new vistas. Understanding the experiences of others is a cognitive event that requires taking into account the beliefs and desires of others, and perhaps seeing them as intentional in the context of action. Intentionality is a cognitive event, and there is a wide diversity of cognitive functions in social reasoning (e.g., Sabini 1992; Kunda 1999).

Figure 2.1
Pondering others (Yansen and Schulkin, unpublished).

Let's trace some of the changes in psychology that set the stage for the ready acceptance of this theory of mind. I want to ensure that this vital capacity is linked to the cognitive ability not only to understand the beliefs and desires of others but to understand other people's experiences. Perhaps we can reinsert the pragmatists' emphasis on experience. Many of us have understood that the aim of psychology as an intellectual pursuit is to understand the human and animal mind and experience. The study of mind and experience is the integrative link that binds all psychological inquiry together.

The tradition of William James (1890/1952, 1912/1958) and John Dewey represents within psychology a link to experience that is construed in broad functionalistic and experiential terms. Dewey (1910, 1925/1989), for example, constantly referred to our adaptations to the world—our problem solving knotted to the search for stability and the experience of security. This view is inherently an evolutionary biological view (Marler and Hamilton 1966). Both James (1890/1952) and Dewey (1910/1965) were responsive to the biological sciences and their methods, and this responsiveness would remain constant throughout their lives. As Dewey (1910/1965) expressed it, "Doubtless the greatest dissolvent in contemporary thought of old questions, the greatest precipitant of new methods, new intentions, new problems, is the one effected by the scientific revolution that found its climax in *The Origin of Species*" (19).

For both James and Dewey or for George Herbert Mead (1934), experience is not simply about sensations (private), but about the experience of objects we encounter in the world in trying to cope and adapt. They viewed experience in the context of transactions with others—negotiating, laboring, loving, and so on. For us, the recognition of the experiences of others is essentially linked to the social world, the world of shared meaning and social practice. Experience, for pragmatists such as James, is a means to engage the world and not something that cuts us off from the world (chapter 1). This sense of experience underlies social reason. The knowledge of each other's experiences is shared and active, not isolated and passive.

As James (1912/1958) said, "knowledge of sensible realities comes to life inside the tissue of experience. It is made and made by relations that unroll in time" (57). And as Dewey (1925/1989) stated, "Experience is not a veil that shuts man off from nature; it is a means of penetrating continually further into the heart of nature" (from the preface). It is a means of coping with nature, of surviving. Any account of mind and experience in either the neural or cognitive sciences, I suggest, needs to recapture this tradition. What makes psychology a unique discipline is its link to experience. For pragmatists such as James or Dewey, mind and experience are tied to function and mechanism (see, for example, James 1907/1959; Dewey 1910). Experience is replete with cognition and embodied in central states of the brain (see chapter 6). It is active and not just passive; for pragmatists, it is linked to action, to the resolution of problems.

Cognitive Science

Behaviorism and its decline gave rise to the cognitive sciences, but it contributed to the development of the art of experimental design. The search for scientific respectability no doubt contributed to the rise of behaviorism (Pavlov 1928), but behaviorism's demise perhaps could have been predicted. Narrow behaviorism was critiqued by Tolman (1932), Konorski (1967), and Miller (1971) from within learning theory. Despite the dominance of behaviorism, many students of behavior eschewed it (e.g., Lashley 1938; Kohler 1925; Hebb 1949) or were impervious to it (e.g., Richter 1942; Griffin 1958; Kluver 1933). But it was Chomsky's (1959) critique of Skinner's book on learning language (Skinner 1957) that helped usher in the age of cognitive science. Chomsky (1972; see Pinker 1994) postulated that language is part of psychology. Language came to be associated with psychology and with the formal properties of mind (e.g., syntax in Chomsky's case). More importantly, psychology was being rescued from the impoverished perspectives of narrow learning theories. After Chomsky, it was once again legitimate to talk about mind. With the work of Miller, Galanter, and Pribram (1960), in addition to that of Simon (1974) and Newell (1990) and many others, cognitive science was born, and behaviorism declined. Many thinkers within philosophy had accepted that cognition pervaded the mind and experience. Still, at places such as Harvard during the 1960s, when students asked about mind, Skinner would send them to Quine in philosophy. When they asked Quine, he would send them back to Skinner. The question was never answered satisfactorily (George Rey, personal communication 1975).

Even in the infancy of cognitive science, skeptics about its approach abounded. For example, Quine (1959), a sophisticated philosophical behaviorist, believed there was no mind. But for the burgeoning cognitive psychology or cognitive science, everything centered on the formal properties of mind. Soon, even Pavlovian conditioning was construed in cognitive terms, as reflective of psychological events such as expecting, drawing inferences, and predicting events (e.g., Rescorla and Wagner 1972; Mackintosh 1975, Dickinson 1980). It was no longer wrong to refer to mind. Consider a quote from one prominent contemporary learning theorist: "for the learning theorist the conditioning experiment is primarily an analytical tractable tool for studying the cognitive changes that take place during learning" (Dickinson 1980, from his preface; see also Dickinson and Shanks 1995).

Learning theory, oddly enough, was reinvigorated by two domains of inquiry: cognitive science and the neural sciences. The Rescorla-Wagner Model (1972), for example, found application in parallel processing systems as models of mind (e.g., McClelland and Rumelhart 1986) and in the neural sciences (e.g., Kandel 1976). If Pavlovian principles of conditioning can become cognitive, we should not be surprised by the power of the trend toward cognitive science. Pavlovian conditioning found its niche in the broader field of information processing (e.g., Rescorla and Wagner 1972), which has been an essential component of the cognitive sciences. The Skinnerian tradition was perhaps more about environmental niches, individual differences, and connectionistic integration in behavior, but in some contexts it was nevertheless knotted to an analysis of innate quality space (Quine 1959) and constraints on foraging behavior and learning (see review by Robinson and Woodward 1989).

Cognitive psychology did not begin with Neisser's 1967 book, but that book solidified what was occurring. Decision sciences were spawned from other avenues of cognitive science, including considerations of the range of information that can be processed and remembered (e.g., Sperling 1960; Sternberg 1969; Glanzer 1968). Investigations exploring how people make decisions were inspired by the study of the psychology of human errors in the work of Kahneman and Tversky (1973; see also Baron 1997). Pure reason was replaced with reason that is imperfect but adequate to the task at hand (Simon 1982) or imperfect reason. A recognition of imperfect reason reinvigorates the pragmatist's position about problem solving and its basis in experience and social knowledge. Moreover, the emphasis on problem solving is tied to biological selection and cultural proclivities. What James (1907/1959) and Dewey (1910, 1910/1965) emphasized were the functionalistic experiences involved in problem solving (see also Peirce 1877, 1889/1992).

Thus, cognitive science is now entrenched in the study of learning, perception, language, and most other branches of traditional psychology. By *cognitive,* I mean processes that are linked to interpretation, memory, anticipation, and problem solving. These processes are often not conscious or controllable (Fodor 1983). Cognitive science often studies the field of perception, sensations, and illusions. Attention is cognitive whether in humans (e.g., Treisman 1964; Posner et al. 1987; Pashler 1998) or in rats and has to do with allocation of mental resources (Mackintosh 1975). In addition, developmental psychology, whether the older form in Piaget (1952) or the more recent work (e.g., Carey 1985), is essentially cognitive. No single definition of the term *cognitive* encompasses all forms (e.g., Parrott and Schulkin 1993; Shettleworth 1999; McFarland 1991).

The overwhelming dominance of cognitive science in the study of memory, perception, emotion, and social behavior, in addition to language and learning, indicates the power of the cognitive turn. These cognitive events are construed as largely unconscious (Fodor 1983; though see Ericsson and Simon 1980 for a defense of some limited transparency; also see Clark and Squire 1998). This orientation to the cognitive did not return to James's notion that psychology, in addition to the study of mental events, also tries to capture the richness of experience. In other words, the rich sense of experience that permeates the James tradition has not been developed within cognitive science (e.g., see Dennett 1969, 1991).

The cognitive revolution needs anchoring with an emphasis on the expression of experience and the biological context in which brains evolved (Preuss 1994). Early on, Dewey (1896) challenged stimulus-response theories of behavior and introduced cognitivism in understanding how we cope with and adapt to our surroundings. But both Dewey (1925/1989) and James wanted to capture the richness of experience linked to function and biological adaption and to behavioral functions in appetitive and consummatory behaviors. Experience is the conduit to get outside one's "head" and be responsive to the world (chapter 1). Any theory that captures humans within the cognitive and neural sciences should try to take that into account.

From Physiological Psychology to the Neural Sciences

An emphasis on the neural and biological sciences within psychology has its roots in James (1890/1952) in the United States. The rupture from biological thinking occurred after James (e.g., Watson 1914; Skinner 1938).

Physiological psychology, in the United States in the twentieth century was heralded by Lashley (1938). Behavior became tied to anatomy. Physiology was being uncovered within Cannon's work (1915/1963) and was linked to behavior (Richter 1942). Morgan wrote the first modern textbook on the subject and the second edition with Stellar in 1954. Hebb (1949) and Beach (1942) were among the first explorers of what became known as physiological psychology, though there is an older tradition that can be traced to the nineteenth century (e.g., see review by Milner and White 1987).

The physiological and anatomical roots expanded dramatically as the field ignited. Sperry (1961) outlined the neural circuits established for behavior in development and then later in split-brain preparations (e.g., Gazzaniga, Bogen, and Sperry 1962). Olds and Milner (1954) discerned the substrates of reward as Beach (1942) described the ethological roots of reproduction. But perhaps two figures stand out as most enhancing the appeal of physiological psychology: Miller (1959) and Stellar (1954). Each was interested in motivation, and before long, the physiological psychology of motivation was being pursued in a number of laboratories. Two major papers expressed this trend. The first, by one of my own professors and senior colleagues, Eliot Stellar, was titled "The Physiology of Motivation" (1954) and the other was Miller's "Chemical Coding in the Brain" (1959). The characterization of motivated behaviors was pursued by ethologists such as Tinbergen (1959; see also Lorenz 1965 and Hinde 1970). Motivated behaviors such as fear or hunger (desires and beliefs about where they can be satisfied) are knotted to regions of the brain that underlie cognitive representation of goal objects (Gallistel 1990) and motor control (e.g., Swanson and Mogenson 1971; Rolls and Treves 1998; see chapter 6).

Physiological psychology (or psychobiology) soon evolved into behavioral neuroscience. As a graduate student in the late 1970s, I had several colleagues who were loath to call themselves physiological psychologists because they considered it a dead field. The new field then was behavioral neuroscience. Before long, one was called simply a neural scientist; behavior was undermined, and the larger sense of experience that James and Dewey had in mind was bypassed. It is one thing to reduce events to simplify them and quite another to reduce psychology out of the picture (see Teitelbaum and Pellis 1992). Psychobiology or behavioral neuroscience should be essentially integrative, respectful to different levels of explanation, and not simply reductionistic (Dewsbury 1991; Lederhendler and Schulkin, n.d.). The pragmatist's orientation toward linking biological and behavioral functions is also integrative and has a rich notion of experience. Again, the pragmatist emphasizes experience as the means by which one engages the world; in other words, one cognitive mechanism for understanding others is to see them as intentional agents.

Modern cognitive neuroscience is a combination of cognitive psychology, computer science, and neuroscience (e.g., Kosslyn and Anderson 1992; Gazzaniga 1995; Posner 1989; Geschwind 1980). With the onset of the modern technical advances in brain imaging (chapter 1), we are now in a position to ask questions about neural circuits and mechanisms that underlie cognitive events such as the discernment of beliefs and desires of others (chapters 5 and 6). These advances have expanded considerably the scope of behavioral neuroscience. Still, no technique replaces good theory both at a behavioral and neural level of analysis.

Capturing the Experiences of Others

The importance of understanding the experiences of others can be found in the idea that the attribution of intentions to others is not just epistemological; there is something about wanting to understand the experience of others. It is part of how we obtain an orientation to our social circumstances. The James-Deweyian view of psychology attempts to discern the features of the environment in which we evolved, as well as the underlying psychobiology, our social evolution, and the utter profundity of the social discourse that determines the content of our thoughts (Mead 1934; Sabini and Schulkin 1994).

But for many of us in the behavioral neural sciences, the mechanisms are unconscious, and thus there is little that is personal in what makes our mental life and our experiences possible (Rey 1999). Trying to reconcile that idea with a rich sense of active experience may seem contradictory, but it is not. I am not talking about sensations or qualia (see Sellars 1963, 1997). I am largely talking about action and adaptation, and how cognition is ubiquitous. In chapter 6, I say more about this subject and about some of the neural sites that underlie social responsiveness.

Two senses of functionalism therefore reinforce one another; the first sense is the one Darwin and the pragmatists held—mind adapting to niches and solving problems (Parrott and Schulkin 1993a). The second sense is tied to the cognitive revolution, the internal computational mechanisms (Putnam 1997; Rey 1997), or what Peirce (1887) meant by "logical machines." Neither sense is strictly reducible to the neural sciences, though emanating from the brain. We should preserve levels of analysis and in fact be respectful of various phenomena (e.g., behavioral, physiological, neural, molecular). The difficult part is bridging across different levels of explanation.

Both senses of functionalism are tied to experience but recognize that experience is not the same as consciousness or learning (e.g., Fodor 1983; Berridge 1999). Moreover, recognizing experiences reflects an ability to parse our social space into beliefs and desires.

Conclusion

The cognitive revolution pervades psychology and neuroscience (see Johnson and Erneling 1997). There is no simple seeing.

All seeing, all sensory or higher-level information processing, comes from a background framework of cognition (Sellars 1963, 1997; Quine 1966) characterized in term of degrees of theory.

The brain is essentially a cognitive organ. Cognition is used for information processing involved in interpretation, memory, anticipation of events, and problem solving (Parrott and Schulkin 1993a), but something can be adaptive without it being cognitive (e.g., white blood cell count in response to infection). Moreover, all cognition is clearly not the same, nor are the regions of the brain that underlie their expression. For example, regions of the brain that mediate declarative (explicit) memory are distinct from those that mediate implicit memory (e.g., Squire and Zola 1996; Mishkin et al. 1997). Interestingly, regions of the brain (the striatum) typically linked to motor control underlie habit formation (Knowlton, Mangels, and Squire 1996), artificial grammar learning, and probabilistic reasoning (Squire and Zola 1996). Perhaps these same systems are linked to learning about others (Lieberman 2000). One question is: How penetrable are these cognitive systems to one another (e.g., Rozin 1976; Fodor 1983; Reber, Knowlton, and Squire 1996; Pylyshyn 1999)? And that is clearly an empirical question.

A set of mechanisms perhaps evolved by natural selection in our mind-brain to parse our social world in order to orchestrate adaptive judgment in action. The social world contains inherent forms of organization (Fiske 1991; Shweder 1991) embodied in common practices (Bourdieu 1972) and in physical/mindful interactions (Lakoff and Johnson 1999), but I know your beliefs and desires as if they are embodied in your experiences. This knowledge of the other's experiences is inherently social (Mead 1932). The structure of life world for us is the social (Schutz 1967). An issue is how to capture this structure in the neural and cognitive sciences.

Let's turn now to consider some of the empirical research about the mind of several primates.

Chapter 3
Experiments on Social Reasoning in Primates

There is a long but cautious tradition of attributing a rich mental life to animals (e.g., Darwin 1872; Romanes 1882; Tolman 1932, 1948). At times, perhaps, that mental life has been too rich and at other times too barren. Darwin clearly reasoned about the continuity of mind in nature; we were not the only ones with a mind. The question is, Who else has a mind and to what extent?

Animal life is rich in cognition or information-processing systems in the mind-brain, but that does not necessarily add up to attributing intentions to others. At least, it may be difficult to demonstrate such intentions in laboratory contexts. In this chapter, I review some of the experiments on social cognition in primates. In the laboratory, the experiments on whether apes have a theory of mind are equivocal. Nonetheless, information-processing systems in the brain underlie the ability to respond, predict, understand, and manipulate the behaviors of others. I end the chapter depicting some of the neurophysiological findings on monkey face and object-trajectory perception.

Say that you believe something (ice cream) is to be found in a particular location, and I know you desire the ice cream. I therefore predict you will go to that place. This is an example of the attribution of a belief and desire to others. Similar experiments are commonplace in determining whether kids or higher primates attribute beliefs and desires to others. It is this ability to attribute beliefs and desires that underlies the notion of theory of mind.

I now turn to some experiments and some issues that underlie our thinking about social reasoning in higher primates. Figure 3.1 depicts four familiar primates.

Figure 3.1
Depiction of four great apes (Yansen and Schulkin, unpublished).

Theory of Mind in Higher Primates

One important paper that emerged toward the end of the 1970s was Premack's and Woodruff's "Do Chimpanzees Have a Theory of Mind?" (1978; see also Woodruff and Premack 1979). Their thesis was simple: for animals or entities to be considered as having a functional state of theory of mind, they must have the ability to attribute beliefs and desires to others, and they would attribute beliefs and desires in the prediction of behavior. This parsing rule may have evolved to predict the behavior of conspecifics. Again, I would suggest that we disassociate the ability to understand other people's experiences from consciousness. Premack, I think, equivocated or was not clear about this distinction.

One part of the issue, as it was originally framed by Premack and his colleagues, is whether chimpanzees *(Pan troglodytes)* had intentions. The second part is whether chimpanzees attribute intentions to others. Premack trained numerous chimpanzees and taught them a nonverbal language. Among his most successful "students" was Sarah, a chimp he acquired when she was quite young.

The experiments were elegant, simple, and, in my view, representative of the experimental legacy of behaviorist psychology (chapter 2) in that they followed the the logic of experimental design. After all, theory was not the legacy of the behaviorists because they eschewed theory for the most part; nor was the conception of the mind their legacy because they eschewed that also. What the behaviorists offered was the logic of experimental design. What better person to begin the demythologization of mental terms than someone within that tradition of parsimony and elegant design? Methodological behaviorism, not ontological behaviorism, is a wonderful restraint on and tool for our mentalistic predilections (see also Heyes and Dickinson 1990; Heyes 1998). Premack's experiments are now well known and quite beautiful.

To test whether chimps demonstrate intentional behavior, Premack and his colleagues designed a number of experiments in which the subjects had an incentive to lie (Woodruff and Premack 1979; Premack and Premack 1983). Consider one example. The subjects, four young chimps behind a partition, were shown a container filled with food, which was then sealed, and an empty container, which was then sealed. A trainer who did not know which container had food came into the room and tried to determine by the animal's actions which container had food. The "friendly" trainer would share the food he found with the chimp; the "unfriendly" trainer would hoard all the food for himself. The situation was then reversed, and the chimps had to guess which container had food. The friendly trainer would try to direct the chimps to the food, whereas the unfriendly trainer would direct them to the empty container.

In the earlier experiments, all the chimps gestured toward the food-filled container out of a natural response to their own knowledge of food. Eventually, all four learned to suppress this response, the first step toward lying. Two began misdirecting the unfriendly trainer to the empty container. One of those two continued to be misled by the unfriendly trainer when he directed her toward the empty container. She could mislead, but apparently she could not understand that the unfriendly trainer was consistently misleading to her. Another chimp behaved in the opposite manner: he refused to be misled by the unfriendly trainer into selecting the empty container, but would not deliberately misdirect that trainer. One chimp could do both: lie and understand deception. According to Premack and his colleagues, the ability to suppress natural reactions, what Premack called the precursor to lying, and the ability to lie deliberately demonstrate that chimps have intentionality. (See figure 3.2.)

Moving on to the next level of intentionality, Premack and his colleagues wanted to learn whether chimps are able to attribute intentions (beliefs and desires) to others. Premack showed Sarah videotapes of human actors trying to get food that was just out of their reach. Sarah was then given photographs showing solutions to the various situations. For example, in a video, an actor is seen jumping up and down to reach a bunch of bananas suspended from above, but one of the photographs shows the actor standing on a chair and grabbing the bananas. Sarah was also

Figure 3.2
In this series of pictures, Keith, an actor whom Sarah liked, demonstrates both good and bad solutions to the inaccessible food problems. Sarah chose all the good solutions for the actor she liked, bad ones for the actor she disliked (Premack and Premack 1983).

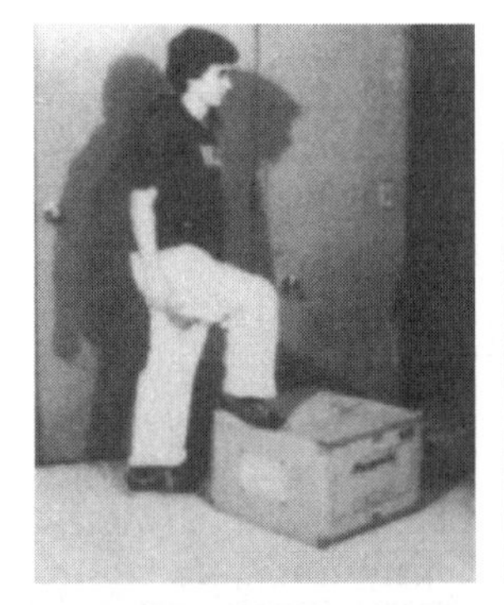

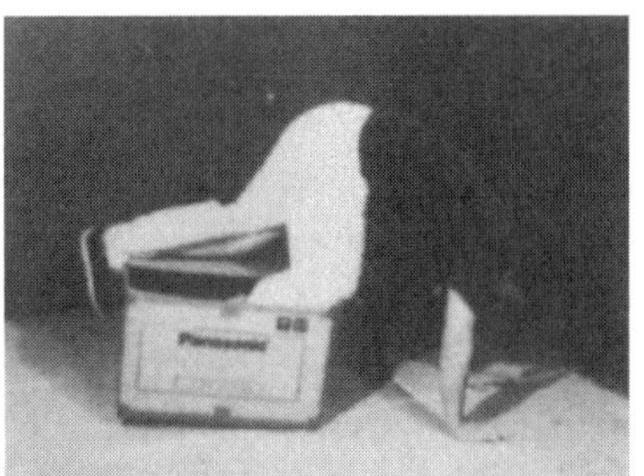

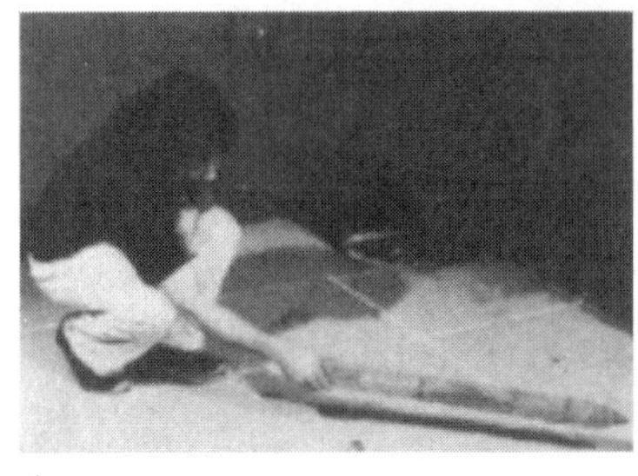

A

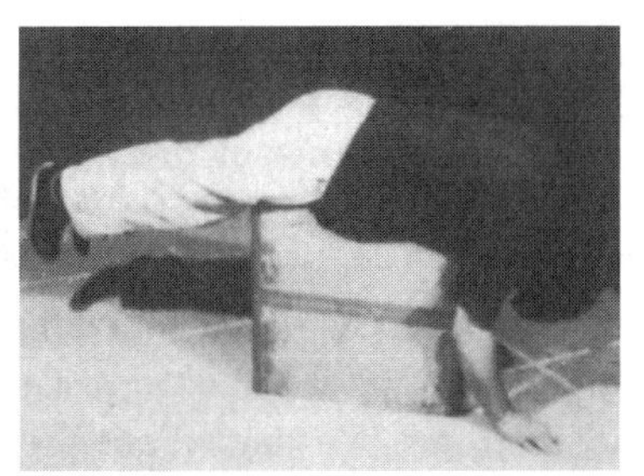

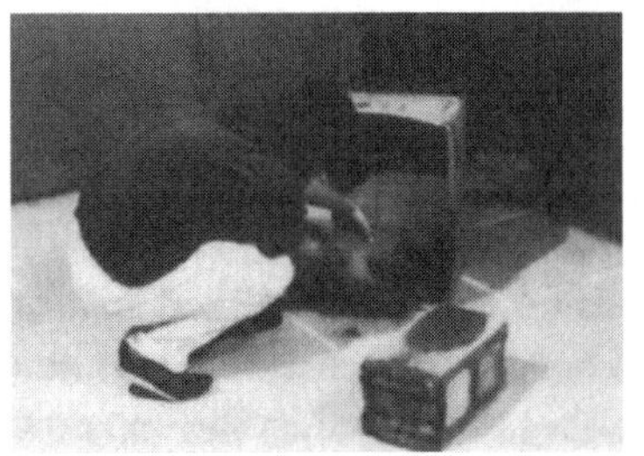

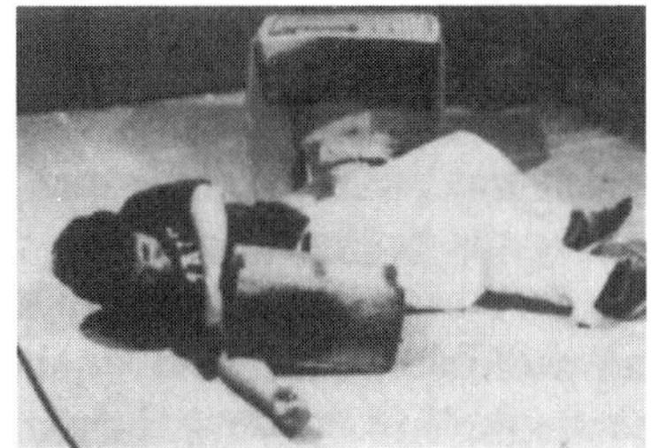

B

shown a picture of a response that would not work in solving the problem.

These experiments paralleled Kohler's classic studies (1925). As you may recall, the Gestalt psychologist, not under the spell of behavioristic constraints, was interested in whether chimpanzees had insight: Do chimpanzees have ideas, and is their behavior more than trial and error? Of course, now we may think a species such as a chimpanzee hit on what Peirce called "abduction," but then so do many other species much less pronounced on the phylogenetic scale (Marler 1961, 1976). What Kohler did was to provide a clear context in which a banana was out of reach. A chair would facilitate the attainment of the goal object. The result was clear. The animal would use a tool to achieve its goal. We know now that tool use is found in many species (Byrne 1995).

Sarah was also given some more complex problems to solve that were not related to food. She saw videotapes of an actor struggling to get out of a locked cage, shivering in the presence of an unlighted heater, trying to play an unplugged record player, and trying to wash the floor with a hose that was not connected to a faucet. In all but two of the situations, Sarah chose the correct solution. Premack noted that Sarah's ability to identify the solution lay in the fact that she recognized the problem. Even though she was not the center of the action, she understood that the actor wanted the bananas, wanted to get out of the cage, wanted the heater on, and so on. Sarah recognized that the videotapes showed not just a sequence of events, but a problem faced by the actor. Problems are basically unrealized intentions. Three-and-a-half-year-old children failed at this experiment, however, choosing pictures that were not solutions but simply related to the objects in the scene. Sarah was sixteen years old when the tests were performed.

Further, Premack asked what question Sarah was answering when she chose a solution: What *will* the actor do? What *should* the actor do? What would *I like* to see him do? He then repeated the experiments using two different actors in each situation, one of whom Sarah liked and one she disliked. She was offered

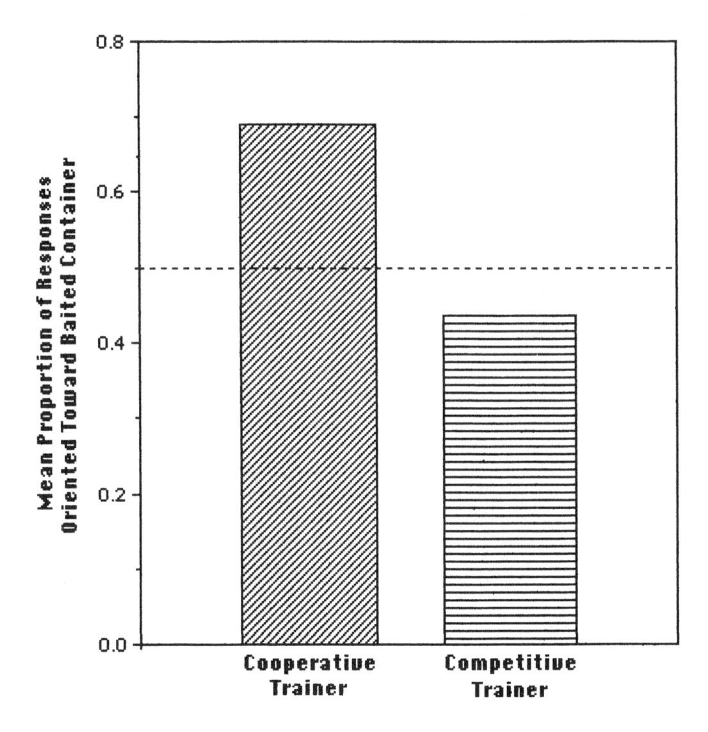

Figure 3.3
Mean proportion of responses toward baited container (adapted from Woodruff and Premack).

pictures that demonstrated a solution, a minor mishap, or an outcome that was both irrelevant and bad. It became clear that Sarah made her choices based on what she would like to see the actor do because she cooperated with the trainer and chose the correct solution for the actor she liked but the minor mishap for the one she disliked. She almost never chose an irrelevant outcome picture. (See figure 3.3.)

The third level of intentionality required that Sarah think that others attribute intentions to her. To test this, Premack showed Sarah videotapes of another chimp, Gussie, taking part in the same experiment she had: watching the actor and selecting the next step from a series of photos. As Gussie was given the photos from which to choose, the videotape Sarah was watching was stopped, and Sarah was given photos of Gussie selecting photos. To demonstrate attribution of intention, Sarah would have had to attribute to Gussie the ability to attribute intentions to the actor in the videotape, which Sarah failed to do.

The Premack experiments had the relevant controls (physical dissimilarities), reflecting the legacy of behaviorism: clear experiment design, use of controls, and so on. What Premack repeatedly demonstrated was cognitive performance that surpassed what we knew about nonhuman primates (e.g., causal knowledge). He suggested a way in which we could ask a close animal relative whether they can do anything like what we do. We will see at the end of the chapter that the issue is far from settled about whether the chimpanzee understood something about the beliefs and desires of others.

The language training Sarah received never did generate syntactic competence, nor did the training she received from other experimentalists (Terrace et al. 1979) generate language competence in any sense like our own, although there still remains disagreement on this issue (e.g., Savage-Rumbaugh 1986; Savage-Rumbaugh and Lewin 1994; Savage-Rumbaugh, Shanker, and Taylor 1998; Sebeok and Umiker-Sebeok (1980); Marler 1999; Gardner and Gardner 1971). What the training apparently did in the Premack experiments was to engender greater cognitive performance on a wide range of tasks (e.g., causal analysis), perhaps including those that require attributing beliefs and desires to others (see also Hayes and Hayes 1951; Marler, in press).

But the luxury of elegant experimental design has its limits—in particular, the confines of artificial space and place. Cognitive competence and intentional attribution may be richly expressed in biological contexts in which these animals are foraging and surviving.

One caveat. Premack was still battling the positivism and behaviorism he learned at the University of Minnesota in graduate school. Theory as he understood it was of the unobservable (e.g., mental events) as opposed to the observable (physical events), which is how he originally described the thesis in the "Theory of Mind" paper in 1978 with Woodruff (see also Premack 1995). This view is a traditional one. Physical events are real; mental events are constructed. Of course, the seeing of physical events is also constructed; the seeing of objects presupposes theory. But Premack did not understand this point, or at least he did not consider it. What he was fighting against was behaviorism; something was mental when the senses could not account for the event.

Nonetheless, Premack provided a laboratory context for considering whether an animal is intentional or not. He concluded that perhaps just apes are. Moreover, he acknowledged that evolved social animals do not necessarily possess a theory of mind (Premack 1995). Let's look at a primate who has diverse social knowledge, but perhaps does not attribute beliefs and desires to others, then we will return to the chimpanzee.

Vervet Monkeys: Communicative Competence and Meaning

Communication pervades the animal's landscape (Smith 1978). Peirce used the term *semiotics,* which became the study of communicative social competence (Habermas 1990).[4] Ethologists and certainly a wide array of naturalists have often referred to the rich mental life of monkeys, which leads to the questions: How do monkeys see the world? What is it like to be a monkey? The answers are perhaps nowhere more clearly expressed than in the work of Cheney and Seyfarth in their book *How Monkeys See the World* (1990b). The title is a bit misleading. The authors are mainly referring to vervets, although they frequently compare and contrast their findings to those about other primates as well as nonprimates. Nonetheless, they have uncovered a rich world of communicative competence. At the heart of their depiction is a wide array of referential capacities. Whereas Premack's work falls into the category of experimental psychology, their line of work is part of the cognitive revolution and falls into the category of ethology—in fact, a new dimension of the category called cognitive ethology (Allen and Bekoff 1997).

African vervet monkeys *(Cercopithecus aethiops)* are a species of old-world monkeys that diverged from humans perhaps some twenty million years ago, in contrast to a divergence with chimpanzees and other greater apes that did not occur until about seven million years ago (Cheney and Seyfarth 1990b).

Vervets can be seen on the jungle floor and in the trees, and are known to inhabit a wide habitat; they are adaptable to many environments. Cheney's and Seyfarth's research took place in Amboseli National Park in Kenya. It is important to note that the setting was the vervet's, not ours.

Vervets are social animals that live in groups of about eleven. Females are stable members of the group, whereas males migrate to other groups when they reach sexual maturity. Social bonds are the backbone of the troop. Social structure is elaborate, and the lineage is maternal, with a few dominant adult males in each group. Figures 3.4 and 3.5 depict an alliance of females against an adult male.

Vervets build and maintain alliances among themselves through behaviors, such as grooming. This social behavior is also used to form alliances based on rank. Females seek to form alliances with high-ranking individuals. The acknowledgment of social rank determines action, as Cheney and Seyfarth demonstrated in the context of unrelated members.

The authors maintain that monkeys recognize other individuals, even those with whom they have little interaction; that they keep track of interactions with others; and that they use that information to influence future interactions. The monkeys also seem to understand the order of social rank within their community and form alliances, which are an important affilitative set of behaviors. Reciprocal social behavior, such as grooming, is inherent in these relationships, and this reciprocity extends to the grooming of each other's infants.

That the vervets keep score of past interactions and act accordingly is evidenced by one of the experiments Cheney and Seyfarth undertook. They observed vervet A grooming vervet B.

Figure 3.4
Two females forming an alliance against an adult male (from Cheney and Sefarth 1990b).

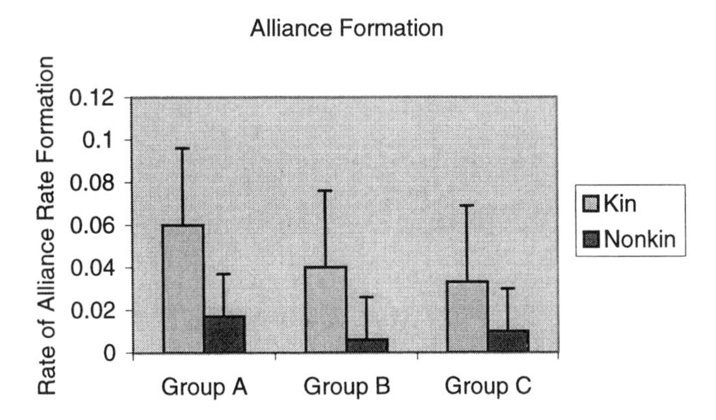

Figure 3.5
Rates at which adult females in three vervet groups formed alliances with each of their kin (light histograms) and nonkin (dark histograms) from three groups (after Cheney and Seyfarth 1990b).

Thirty to ninety minutes later, they played a recording of A's "threat grunt" within earshot of B. A threat grunt is a vocalization a vervet gives when threatening or chasing another member; it also appears to be a call for support from allies. Because A had groomed B, B should have responded to A's threat grunt (in the experiment, looking in the direction of the speaker was considered a response). The researchers repeated the experiment on days when A had not groomed B. They tested pairs of unrelated vervets, as well as mother-offspring pairs and sibling pairs.

Among related animal pairs, prior grooming had no effect on B's response to A's threat grunt. Among unrelated pairs, however, prior grooming strongly affected response, in some cases more strongly than the responses of related pairs. As Cheney and Seyfarth put it, the monkey considers two things: "Is the animal a relative?" and, if not, "What has she done for me lately?" The authors point out that actual physical involvement on behalf of an ally calling for aid depends much on social ranking and the perceived costs of intervention (Cheney and Seyfarth 1990a, b).

Other observations demonstrated that the monkeys recognize the relationships between others, specifically observations of redirected aggression. An animal that has been involved in a fight will sometimes threaten an "innocent bystander," another of the group that was not involved in the fight. Cheney and Seyfarth cite earlier research showing that the previously uninvolved individual is usually a close relative or even a close (but unrelated) friend of the opponent (Cheney and Seyfarth 1986, 1989).

On the flip side, reconciliation often follows the same pattern as redirected aggression (Cheney and Seyfarth 1990). After a fight, monkeys may reconcile by touching, hugging, or grooming the opponent or even by handling the opponent's infant. They also reconcile with the relatives of the opponent in the same manner.

Cheney and Seyfarth attribute much cognitive competence to the vervets. They suggest that the male assess the closeness of bonds between pairs not only in their own group but also in other troops of monkeys (Cheney and Seyfarth 1990). They cite an experiment in which rival monkeys observed the dominant males with their female companions. The lower-ranking rivals apparently chose not to challenge the dominant males for females with whom the dominant male had a very close bond (as evidenced by the amount of time the female spent grooming the male). The rival males evidently assessed the closeness of relationships, which informed their behavior.

Cheney and Seyfarth tested whether the vervets' various calls could be evaluated in terms of meaning and intention (Cheney and Seyfarth 1990). They began with Dennett's (1969; 1987; 1996) concepts on levels of intentionality (Dennett; see Cheney and Seyfarth 1990). Zero-order intentional systems have no beliefs or desires but rather emit a response in reaction to a stimulus. First-order intentional systems have beliefs and desires but no beliefs about beliefs—that is, they have no concept of the listener's state of mind. Higher-order (second-order and up) intentional systems are capable of developing concepts of what others are thinking and taking into account both their own and others' beliefs and desires. The authors point out that animals' calls can communicate much about their environment without the caller having any idea of the effect of the call on his audience.

An important point is that vervets emit calls in differential contexts. For example, they have a specific alarm call to signal the presence of eagles. Cheney and Seyfarth note that vervets usually look up when they hear this alarm call, but not always. Sometimes they jump out of trees or run into bushes; sometimes they do not respond at all. Sometimes they respond to hearing an alarm call by giving another alarm call—but again, not always (Seyfarth, Cheney, and Marler 1980). Apparently, vervets can choose their responses, which means that their calls are voluntary and not entirely mechanical responses (Cheney and Seyfarth 1990). Further, solitary vervets do not give alarm calls when confronted by a predator, again demonstrating that vervets choose to give signals based on the circumstances.

The authors went on to study this phenomenon with a group of captive monkeys. In a cage that had both indoor and outdoor areas, a female adult was isolated outdoors with either one of her offspring or an unrelated juvenile of about the same age (the other monkeys of the group were confined indoors). A possible predator (a veterinarian surgeon carrying a net) approached the two monkeys outside. The adult females gave significantly more frequent alarm calls when they were with their offspring than when they were with an unrelated juvenile (Cheney and Seyfarth 1985). When the same experiment was performed with adult males, the males gave more frequent alarm calls when they were with a female monkey than when they were with a dominant male (Cheney and Seyfarth 1985). The authors conclude that vervets modify their calls based on their audience. They point out that other species do the same, including squirrels and downy woodpeckers.

To examine whether vervets can attribute beliefs to other vervets, Cheney and Seyfarth designed an experiment in which a monkey's calls would come to be heard as unreliable—that is, considered false alarms. They would repeatedly play one monkey's alarm call for signaling that a rival vervet group is nearby even though no other vervet group was in the area (Cheney and Seyfarth 1990). The vervets who heard the false alarm did, in fact, begin to ignore the call. They responded, however, when another individual gave the same alarm call. Seemingly, the listeners take into account both the individual source of the call and the reliability of that source in a particular context (Cheney and Seyfarth 1990). Therefore, individual calls take on specific meaning, and that meaning is based on one monkey's (the listener's) beliefs about another monkey's (the caller's) beliefs—that is, the listener may believe that the caller is either deceived or deceiving.

Once we begin to understand the vocabulary of nonhuman animals and the ability to communicate that it implies, the question of reference continues to nag at us. When a vervet monkey gives the eagle alarm call, does the call itself provide sufficient information about the predator for the listener to respond appropriately? Experiments that use recordings to determine how and when a subject will respond were undertaken to evaluate how well a vervet monkey can distinguish a call from its context (Seyfarth, Cheney, and Marler 1980).

When the vervets heard the leopard alarm call, those on the ground leapt into nearby trees, and those in trees climbed higher into the trees. When they heard the eagle alarm call, they looked up into the sky and then ran for cover on the ground. When they heard the snake alarm call, they stood on two legs and scanned the ground around them. Clearly, each call elicited a specific and appropriate response. Some researchers have argued that the different calls merely express the caller's level of fear, but others dispute this argument (Cheney and Seyfarth 1990b; Marler, Evans, and Hauser 1992).

With the ability to communicate comes, apparently, the ability to deceive. Nonhuman animals do in fact engage in deception, and Hauser (1996) distinguishes between functional and intended deception. Deception can involve active falsification or the withholding of information (Hauser 1996), and studies of chickens and monkey demonstrate how and when animals employ these tactics.

The life of the vervet monkey is quite rich without it being intentional in the third-order sense. In fact, Cheyney and Sefarth think that this species is not. Recognizing the constraints on the use of a concept is fundamental, as we all know regarding a concept's legitimacy in our scientific lexicon. Let's consider a bit more of the richness of this species' mental life.

Social Intelligence Pervades Primate Evolution

A prominent hypothesis in the primate literature (Jolly 1966, 1985; Humphrey 1976), one with much warrant, is that social cognition is at the heart of primate evolution. The description in the previous section clearly highlights social intelligence (e.g., Dunbar 1988; Runciman et al. 1996).

What do we know from these sorts of experiments about the mind of the vervet? Clearly they are socially responsive. I would suggest again that the mechanisms that make this responsiveness possible are unconscious and that the monkeys themselves might not be aware that they are making these sort of appraisals of their social milieu.

Cheney and Seyfarth suggest that recognition of social rank and reciprocity is represented in and the recency of cooperation determines the vervets' social interactions. The vervets' use of social knowledge is apparent in the endless prediction of behavior and events in the world to which they are adapting.

It is a familiar thesis that the ability to cope with complex social interactions is an adaptive trait in primates (Jolly 1966, Humphrey 1976). Much of the intelligence primates display in laboratory experiments can be compared with the social intelligence they display in their interactions with others in the wild. For example, experiments of transitive inference may represent the assessment of dominance ranks (Cheney and Seyfarth 1990). This ability in the laboratory also represents the access of a cognitive ability being applied to a novel context, which is a feature in the evolution of intelligence (Rozin 1976).

Cheney and Seyfarth have also compared social intelligence in monkeys with the development of social intelligence in very young children. Human infants, they note, are more attentive to human faces than to visual depictions of inanimate objects, and they are more responsive to speech sounds than to other sounds. Moreover, humans understand causation in a social context sooner than they understand the effects of one inanimate object on another (Fein 1972; Cheney and Seyfarth 1990).

Vervets have rich social cognitive abilities. The same no doubt holds for us. Social cognition is not remotely equal to theory of mind, however. Consider several more facts. In laboratory experiments, monkeys who have learned to press a lever that reveals a picture are more interested in looking at pictures of monkeys of their own species than at other species. They can also distinguish individual monkeys of their own species. The subjects quickly became bored of looking at the same monkey (of their own species), but regained interest when they looked at other individual monkeys (Humphrey 1976; see Cheney and Seyfarth 1990). In contrast, the monkeys also became bored of looking at pictures of pigs, apparently because they did not distinguish even very dissimilar pigs as individuals. Of course, pigs may also be less interesting.

To further evaluate vervets' social knowledge, investigators tested what the monkeys understood about other species' behavior. For example, vervets recognize the alarm call that superb starlings give when they see a raptor (Cheney and Seyfarth 1990). To the vervets, however, the sterling's land predator alarm call may signify either a leopard or an eagle. Although the vervets also seem to recognize the alarm call that the impala gives when it sees a leopard, they are unable to comprehend that the starling's predator alarm and the impala predator alarm signal the same thing (Cheney and Seyfarth 1990). They conclude that vervets are understanding the world from only their own point of view (Cheney and Seyfarth 1990).

Vervets can make limited associations across species. For example, the Masai tribesmen bring their livestock to the Amboseli Park to graze during the dry season. When the vervets see a Masai person approach, they give an alarm call specifically for strange human beings. Although the livestock are not threatening to the vervets, one would expect the presence of livestock to signal to the vervets the presence of the Masai and, in such cases, for the vervets to give the strange-human alarm call. Indeed, there is some anecdotal evidence of such an occurrence. Cheney and Seyfarth tested whether vervets could distinguish between the lowing of cattle and the lowing of wildebeests, neither of which animals are threatening to the vervets. In the experiment, the vervets responded with more concern to the sound of the cattle than to the sound of the wildebeests. However, none of the vervets ever responded by giving the strange-human alarm call (Cheney and Seyfarth 1990b).

Some Doubts about the Chimpanzee Work

Let me return to the work of the chimpanzee. I intimated that later investigators did not agree with the Premack work (e.g., Povinelli and Eddy 1996a, b; Tomasello and Call 1997). The paradigms, however, are different. I also think that in all of this work, there is some confusion about consciousness or awareness. It is as if the chimpanzee needs to be aware that they are doing something in order to instantiate this theory of mind, or at least there is equivocation on this conclusion, but I think it is a mistake. Nothing needs to be conscious (in terms of mechanism) in order to attribute beliefs and desires to others.

Let's start with self-recognition. Self-recognition in mirror and other paradigms has been characterized in great apes (Gallup, 1979, 1981; Gallup, Povinelli, and Eddy 1995. It has been suggested that this self-recognition precedes knowledge attribution to events—a suggestion that, of course, hangs on the notion of knowledge.

Facial and bodily recognition do not necessarily mean that chimpanzees has a sense of themselves (Povinneli and Eddy 1996a). Even eye contact—an important factor in joint attention (namely, the two of us both attending to the same object and being able to jointly respond to that object) and in recognition of the experiences of others in us—may not necessarily be significant for theory of mind in chimpanzees (Heyes 1998; Povinelli and Eddy 1996a).

One thread in the literature that may be a source of confusion is that some researchers see eye contact as indeed a way of looking in on someone's experiences (e.g., Baron-Cohen 1996), but eye contact can be maintained without focusing in on someone (Povinelli and Eddy 1996a). You and I may regularly make eye contact without either of us being responsive to the experiences of others, even though we could.

One suggestion: When we move from the mirror self-recognition, which chimpanzees can clearly do, experiments with chimpanzees regarding visual connection with human experimenters reveal that the chimps in part and act out of a behaviorist framework of understanding and not a mentalist framework (Povinelli and Eddy 1996b). The researchers set out to determine whether chimpanzees understand whether visual perception is understood as being connected to the external world. (See figure 3.6.)

One view suggests that chimps do understand the connection (Ristau 1998). In experiments, they would direct a begging-for-food gesture to the experimenter who could see them but not to one who could not see them. Another view suggests that the chimps do not imbue another's vision with any particular intention, and so in experiments they would beg from both experimenters equally. In the latter view, chimps could process information gained about eyes and looking, and eventually use it to determine which experimenter was more likely to produce food in a given situation.

Researchers (Povinelli and Eddy 1996a) capitalized on a group of young (four to five years old) captive chimpanzees' tendency to beg by the gesture of presenting an outstretched hand, palm up, to test the idea of "seeing" as a mental event. They trained the chimps to associate this gesture with a reward of food. In all the experiments, the chimp was on one side of a Plexiglass partition with holes large enough to reach through, and two experimenters were on the other side. In a series of baseline experiments, the chimps learned to perform their begging gesture toward one or the other experimenter.

In the study, the chimps had to choose between an experimenter whose vision was blocked and one whose vision was not. Only the experimenter who could see the chimp would produce food when the chimp directed a begging gesture toward that experimenter. The experimenters blocked their vision by means of wearing a blindfold, putting a bucket over the head, placing hands over the eyes, turning the body completely away (back to the subject) and, in some tests, simply by closing the eyes. The numerous experiments sought to take into account naturalistic cases of blocked vision (eg., hands over the eyes) versus "unnatural" cases (e.g., a bucket over the head), the emotional arousal caused by seeing the experimenter blindfolded, and other such variables.

In the earliest trials, the chimps failed to demonstrate the correct behavior—i.e., gesturing toward the experimenter who could see them—in most of the tests, with the exception of the case in which the experimenter's back was turned. This result seemed to support, to some extent, the mentalist framework of thinking. By later trials, however, it became apparent that the

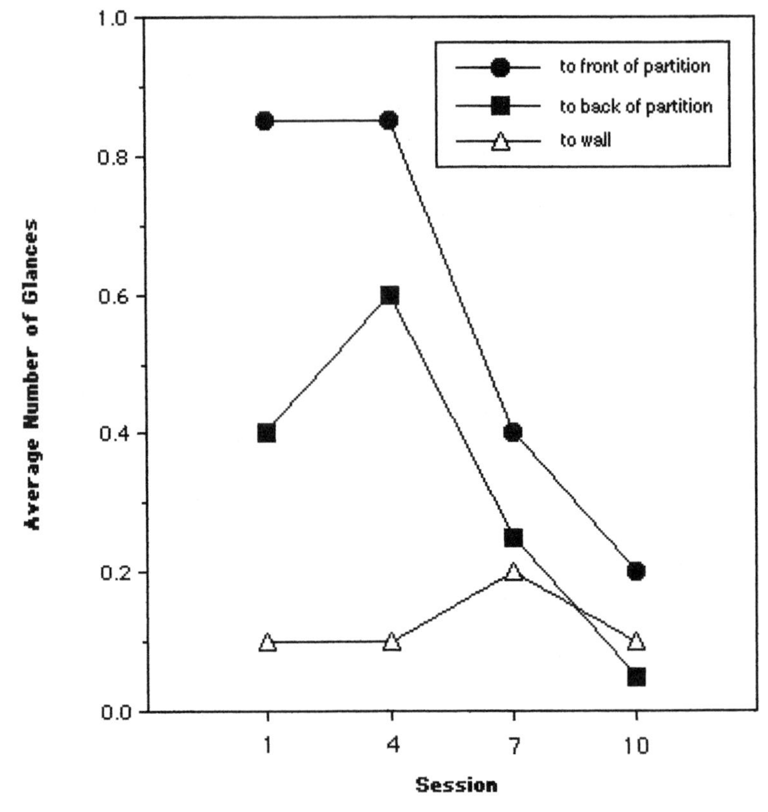

Figure 3.6
Percentage of trials (± SEM) eliciting gaze-following behavior in response to eye and head movements in concert (eyes 43 head), eye movements only (eyes), and no movements (control). Each bar represents the group mean for the seven subjects (Povinelli and Eddy 1996b).

chimps could not, in fact, successfully associate eye gaze with the experimenter who could see them. They had begun to perform more successfully, but the cause appeared to be that, through repeated trials (over a period of months), the chimps had learned to make an association—namely, "gesture toward the person whose face is visible." When the researchers applied this rule of behavior, they correctly predicted the chimps' behavior in six out of seven types of tests. This rule also disqualified some other theories—for example, that the chimps would gesture toward the person who could see them based on the fact that they (the chimps) could see the experimenter's eyes. When faced with two experimenters in the same position, one with eyes closed and one with eyes open, they performed at random, chance levels. The fact that the chimps had no particular preference for eyes-open versus eyes-closed in that specific case seemed to discount the mentalist framework (Povinelli et al. 1995).

Of course, in the experiment I just described, the chimps might not express this ability. On the other hand, Povinelli and his colleagues have not been able to demonstrate theory of mind in the way in which Premack envisioned it. It has been argued that what stands out is that chimpanzees do not see others as minds in anything like what we do (Povineli and Eddy 1996a; Povinelli and Preuss 1995). This no doubt is true.

But then the debate switched to so-called "noncognitive" functions that underlie what the chimpanzee does in a social context. I am not sure what this idea means, except that it resurrects a narrow notion of the behaviorism of the past—that is, it conveys a narrow notion of mind—and is confusing and equivocal (e.g., Povinelli and Eddy 1996a). Surely the idea that social stimuli are operative in joint attention is itself cognitive, and the extent that these events are related to reproductive fitness does not mean that they are not cognitive. In the attempt to refute Premack and others, the pendulum can swing back to a view of the cognitive mind that sounds ancient and outmoded—or perhaps worse, strawmanlike and misleading.

Thus, in general, several groups of researchers have not found in laboratory settings any examples of intentional attributional competence toward others in the chimpanzee, a creature so close to us. If the phenomenon of intentionality were robust, it might be easy to demonstrate. It is not, nor is the understanding of causation (Povinelli and Eddy 1996; Tomasello and Call 1997). Attempts by several other groups of investigators to extend

Premack's findings in chimpanzees, have not yet been promising.

The evidence is merely equivocal, however, not hopeless. One experiment demonstrated that chimpanzees and orangutans (*Pongo pygmarus,* but recall that orangutans are not social animals) can distinguish somewhat when a reward is placed at a location intentionally by the experiment and when it is selected accidentally (Call and Tomasello 1998). Both species preferred the box when the reward was placed intentionally by the experimenter. However one explains this meager performance on this one condition: it pales when compared to the performance of three year old humans (see the next chapter). Interestingly, with regard to the orangutans, the animal reared at home with the trainer was quite a bit better at this task than those animals that were not. (See figure 3.7.)

Tomasello and Call (1997) suggest that understanding something about the intentions of another is prior to understanding the beliefs of other conspecifics in this species. An odd thing to say because to be intentional is already to have beliefs and desires.

The anecdotal observations of many researchers of the chimpanzee regularly find them to be intentional (Goodal 1986; Hayes and Hayes 1951; Savage-Rambaugh 1986) or simply to communicate intentionally with others (e.g., Leavens and Hopkins 1998). The latter context is a long way from understanding the beliefs and desires of others, and then using that knowledge to predict behavior. Moreover, perhaps it might be easier to observe intentional attributions in these species in their natural context.[5]

The study of social interactions suggests that gorillas (*Gorilla gorilla;* Gomez 1990, 1991) appear to have some knowledge of other minds—that is, understand something about the trainers knowledge of an event (Gomez 1996). Gomez, who works with gorillas, has provided some evidence that they can share gaze; what he has in mind here is a young gorilla pushing a trainer to look at something explicitly.

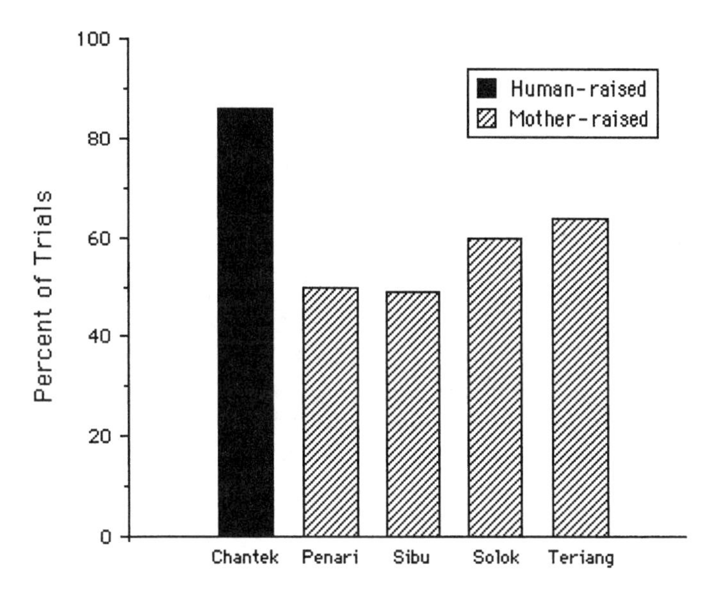

Figure 3.7
Percentage of trials in which individual orangutans differentially selected intentional actions over accidental actions (adapted from Call and Tomasello 1998).

Of course, there are other interpretations of such an occurrence. Gomez (1991) seems to think that the gorilla has what he calls an "implicit theory" or less than a full theory of mind. There is more than one notion of mind; the mind is not a thing, but a collection of organizing principles that underlie behavioral adaptation.

Now whether these species have anything in their cognitive arsenal that amounts to a theory of mind is another question. Moreover, the literature on deception in animals (DeWaal 1989; Byrne and Whiten 1988; Menzel and Juno 1985) has suggested theory of mind in some, but certainly not all instances of sophisticated deception (Byrne 1995). What is involved is misleading others or manipulating others or both to attain a goal; the behavior may or may not involve the recognition of another's mind. Intentional deception and the wars of nerves in everyday discourse (Goffman 1959, 1971; Whiten and Byrne 1988; Whiten 1993) can seem and sometimes are ruinous at the core, but deception goes only so far; eventually one is found out and castigated. Nonetheless, intentional deception is an art—Machiavellian in the pejorative of the great thinker.

Constraints and Backlash

Construing gaze as noncognitive and inferences about mind as cognitive (Povinelli and Eddy 1996; Tomasello and Call 1997) resurrects the old empiricist distinction between sense and form, or sense data and cognition. What we have learned is that sense is always replete with cognition (i.e., information-processing systems); when we speak of the mind, nothing is noncognitive. Gaze is no different. Moreover, the cognitive revolution as I indicated in chapter 2, has transformed traditional behaviorism (e.g., Dickinson 1980; Rescorla and Wagner 1972).

There is little doubt that chimpanzees differentiate themselves from others and notice when something is different about them (Eddy, Gallup, and Povinelli 1996), but they do not obviously direct the gaze of others, though they can use eye-gaze information (Povinelli and Eddy 1996a). They do not seem to use eye gaze to reveal knowing something about the beliefs and desires of others, a fact that is taken to prove that chimpanzees, unlike normal human three year olds, do not have a theory of mind remotely like ours (Povinelli and Eddy 1996a). And of course they probably do not.

Moreover, in the case of causation, having unconscious causal mechanisms and being able to demonstrate knowledge of such mechanisms are very different. There seems to be confusion about this distinction in the animal literature. Again, the mechanisms that render causation and recognition of causation need not be conscious (nor need the recognition as such).

Chimpanzees in the laboratory do not seem to use pointing as young children do (Povinelli and Davis 1994). Also, as I indicated, chimps do gaze in joint visual attention tasks; it is not clear that this indicates that they have a window into the mind or experiences of others (Povinelli and Eddy 1996a). Acknowledging degrees of epistemological opacity, I am prepared to believe that they do, and I think one has to be honest and admit to what is clear and what is not.

Tomasello and his colleagues (1993, 1997) have provided an intentional model of "cultural" learning that includes two features they believe are found only in humans: intentional collaboration and instructual learning. Although Tomasello concedes mental competence in the ability to understand third-party relationships and that this ability is fundamental for social knowledge, that is about as far as he is willing to go at this point. (See table 3.1.)

Tomasello and Call (1997) seem to think that consciousness is required for these sorts of mental competence and concede that if a chimp has a theory of mind, it is not "human like" (400). And of course it is not. Intentional communication, which Tomasello concedes, sounds pretty good to me. Remove consciousness from the picture, as somehow critical, then intentional communication would need to be unpacked, which I am not sure Tomasello does.

On the other hand, the issue is not settled about what constitutes a theory of mind for animals that are close to us. Human infants readily imitate (Meltzoff and Moore 1994) and are instructed in learning in everyday life. The other issue that Tomasello raised for contrasting us with the chimpanzee is complex collaborative learning. But how complex is complex?

Joint attention is endemic to the process when the subjects know that they are both attending to the same events. Knowing turns out to be conscious on many of these accounts and strikes me as restrictive. After all, joint attention can go on without one knowing about it.

Chimps certainly have differences in specific practices (cf. Quiatt and Reynolds 1993/1995; Whiten et al. 1999). They have semantics and social syntax for parsing their rich world and elaborate social space. They can imitate others; they may learn by instruction, which depends on how broadly or narrowly one defines it. Joint attention and collaboration may be

Table 3.1
Tomasello and colleagues' intentional model of cultural learning

Cultural learning process	Social-cognitive ability	Concept of person	Cognitive representation
Imitative (9 months)	Perspective talking (e.g., joint attention, social referencing)	Intentional agent	Simple (other's perspective)
Instructed (4 years)	Intersubjectivity (e.g., false-belief task, intentional deception)	Mental agent	Alternating/ coordinating other's and own perspective
Collaborative (6 years)	Recursive intersubjectivity (e.g., embedded mental-state language	Reflective agent	Integrated (dyad's intersubjectivity)

From Tomasello et al. 1993.

operative on hunts or defense of territory, and they can certainly be characterized as intentional communicators (Leavens and Hopkins 1998). The problem is how to show this intentionality. Most studies in animals where it is attempted do not demonstrate this ability.[6]

Brain, Bodily Motion, and Faces

Of course, a reasonable strategedy is to discern in mammalian nonhuman primates which regions of the brain are fundamental for knowledge—social and otherwise (e.g., Brothers 1994; Baron-Cohen et al. 1999; Hauser 1999). More specifically, one reasonable and tractable question is how the brain responds to different objects (e.g., Desimone et al. 1984; Rolls, Treves, and Tovee 1997). For example, we now know some of the regions of the brain (i.e., temporal cortex) and how these brain regions respond to faces (e.g., Desimone et al. 1984; Gross 1992; Perrett et al. 1998) or to different kinds of object motion (Oram and

Perrett 1996; Hietanen and Perrett 1996). Different visual-processing pathways are thought to mediate the identification of the location of an object, the form of an object, and the direction or the motion and trajectory of an object (Ungerleider and Mishkin 1982; Ungerleider and Haxby 1994). A ventral visual pathway to the temporal lobe is known to represent a number of objects (e.g., faces, houses, chairs, etc. [Ishai et al. 1999]) in the coding of semantics in the brain.

Thus, regions of the temporal cortex compute faces, bodily posture, and motion (e.g., Perrett and Mistlin 1990). Facial and bodily responses are great sources of information (e.g., Darwin 1872; Ekman 1973; Young 1998). Neurons within the temporal lobe are responsive to the individual attention of others, the direction of their eyes, and the direction of their motion (e.g., Perrett et al. 1992; Rolls and Treves 1998). For example, in one study, neurons within the superior temporal sulcus were recorded in macaques. They were responsive to both human and monkey faces (Rolls, Treves, and Tovee 1997) to bodily posture, and motion direction (Perrett et al. 1992). The firing pattern was more significantly elevated when the gaze of another was direct than when it was not. The direct gaze is typically a threatening gesture in nonhuman primates (e.g., Hinde 1970), and therefore this region of the brain is probably linked to the fear information–processing system in the brain (Rolls 1999). (See figures 3.8 and 3.9.)

Summary

Do chimpanzees have a theory of mind? If it is defined as the attribution of beliefs and desires to others in the prediction of their behavior, I would say the evidence is still equivocal (see also Heyes 1998). Partly, this equivocation is definitional. Some still think that such attribution has to be conscious to the animal, but others are not so sure.

The cognitivist position is that information is categorized relative to a background body of knowledge. The issue is not about consciousness; information-processing systems are mostly unconscious. Specialized adaptive functions are key to our

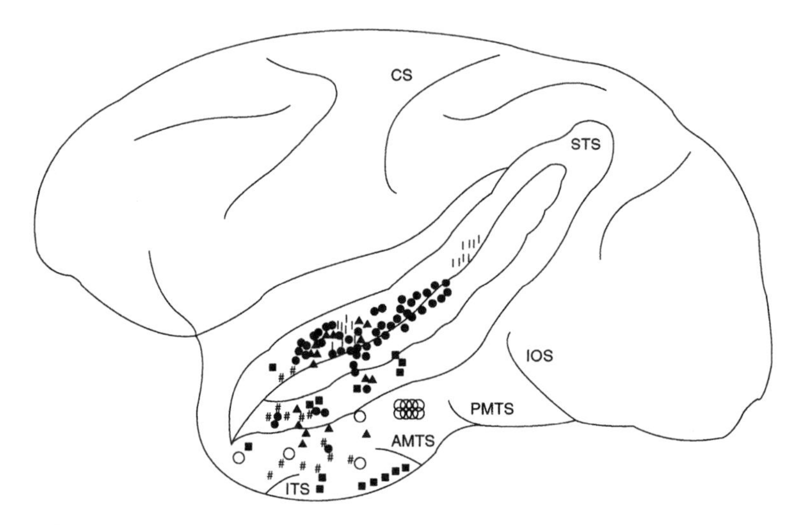

Figure 3.8
Location of cells in the temporal cortex selective for faces from various studies. Drawing of a left side of a rhesus macaque brain showing the major sites. Abbreviations: STS, superior temporal sulcus; IOS, interior occipital sulcus; CS, central sulcus; ITS, interior temporal sulcus; AMTS, anterior medial temporal sulcus; PMTS, posterior medial temporal sulcus (after Perrett et al. 1992).

evolutionary success, and gaining access to specialized functions in the "cognitive unconscious" and extending them in use are part of the evolution of our intelligence (Rozin 1976).[7] The cognitive unconscious underlies the calls of vervets or other animals, as it does for us. The calls are embodied in semantic networks that confer the everyday meaning of survival value. What pervades these species is an information-processing system in the brain, some of which is embodied in action patterns, praxis, and fairly sophisticated mechanisms for social understanding. The formation and maintenance of alliances is no trivial fact (DeWaal 1989). Peacemaking in social relations is a formidable chore, underlying which is perhaps recognition of others. Whether animals really know or not, those others remain.

Although cognitive ethology is an advance over myopic behaviorism in the study of animal behavior (e.g., Hauser 1996;

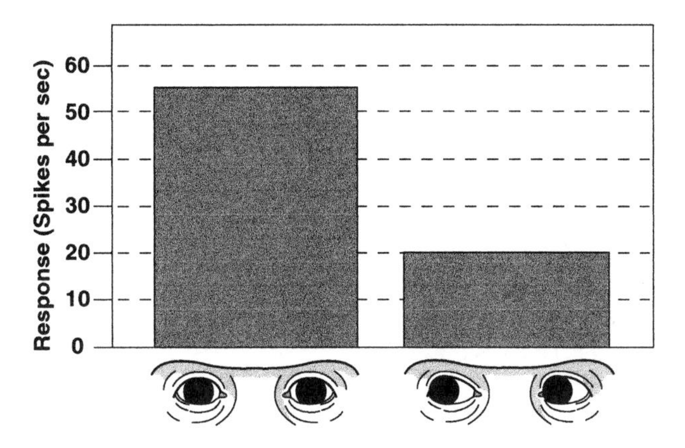

Figure 3.9
Neurophysiological responses from cells in the superior temporal sulcus of the macaque to indirect and direct eye contact of conspecifics (adapted from Perrett and Mistlin 1990).

Shettleworth 1999), many of the questions about other minds remain a puzzle, but we have a pretty good idea that understanding in these species includes both behavioristic and mentalistic attributions. What we need are constraints on the applications of these attributions; otherwise, pernicious behaviorism and endless mentalism loom on the horizon.

What is it like to be someone else or another animal? What are their experiences like? These questions are fundamental to us as human knowers of ourselves and other animals (Nagel 1974; Griffin 1958; Ristau 1998). The theory of mind may have first evolved as something to further parse out social space. It may then have been used to understand and perhaps even appreciate and manipulate the experiences of others. Understanding others' experiences is something we can do fairly easily. In us, this sort of ability may presuppose the intentional stance as a social parser.

Let's turn to children, where it is much easier to discover what they know and how they move from social referencing to reading actions and other minds.

Chapter 4
Development of Social Reason in Children

Conceptions of cognitive development have gone through some radical changes in these last twenty years; Piaget has been challenged on a number of fronts, particularly the idea that we go from sensorimotor to ideation in clear, distinct ways (e.g., Carey 1985; Keil 1992; Lakoff and Johnson 1999). Sensorimotor systems are themselves cognitive (chapter 6; Graybiel 1997; Clark 1996/1997). What has replaced this traditional view is the idea that computational systems or cognitive activity predominates much earlier than Piaget thought (e.g., Gopnik 1993). His view (which is also James's or Freud's view) was traditional to the core: sensations first, ideas second.

But the child's world is no "blooming buzzing confusion," contrary to James's pronouncements; categorical knowledge, not conscious as such, predominates from the beginning (Spelke et al. 1992; Hirschfeld and Gelman 1994/1998). The child's world is never simply a sensorimotor one as Piaget suggested (e.g., Brunner 1975a,b; Gopnik 1993). Psychological categories (ideas about objects, time, probability, and so on) pervade the neonate's world (Carey 1987; Spelke et al. 1992) and underlie action. As Kagan (1984) put it, "The infant's first knowledge is dependent on actions and perceptual experience" (31). The child's world is replete with thought in action (Harris 1996). The young child, as some research has suggested, may be guided by hypotheses and looks for evidence or away from contrary evidence, as the rest of us do (e.g., Koslowski 1996, but also see Harris 1996). The important point is that by three or four years of age the child quite noticeably attributes intentions to others and perhaps begins to cut the social space with the ability.[8]

There is a continuing debate over the extent to which these cognitive systems are modular and penetrable (e.g., Fodor 1983; Karmilof-Smith 1992; Flavell 1999), regarding which I am mostly undecided. These questions are empricial, and not much evidence seems to support the view that early visual information processing is somewhat impenetrable by higher-order cognitive systems (Pylyshyn 1999). Brightness and color illusions do not go away when one knows they are illusions. They still look the same. Is there a social-cognitive equivalent? Perhaps something as important as social cognition should have access to a number of lower-level cognitive systems.

In the rest of this chapter, I present the thesis that part of what underlies social knowledge, in addition to basic cognition, is intentional action being attributed to others, including the concept of animacy. This is one important way in which we form important links to our world. Other people are not just part of the furniture; they are alive. The concept of what is alive and what is not is a fundamental means by which we categorize the world, and it probably reflects our evolution. Intentionality has to do with understanding entities that are alive—their plans. Children express this ability early on. Contrary to a tendency in the literature to separate desires from beliefs concerning theory of mind (e.g., Wellman), however, I suggest that desires presuppose beliefs and information-processing systems in the brain (figure 4.1).

Natural Kinds

Young children have a propensity to reason about natural kinds (Keil 1992; Carey 1985; Gelman and Coley 1990; Johnson and Carey 1998). What this propensity actually reveals is in dispute (Fodor 1994, 1998), but I believe the preparedness to "see" natural kinds is one clear way in which they become anchored to the worlds in which they have to survive (cf. Gibson 1979; Clark 1996/1997; Kornblith 1993).

Clearly some concepts are easily acquired by children. One basic conceptual way to divide the world is into animate and

Figure 4.1
Danielle and Nicholas Schulkin (my then six-year-old and six-month-old).

inanimate objects. More generally, there is a preparedness to learn some things more easily than others (Garcia and Ervin 1968; Seligman 1970; Shettleworth 1971; Rozin 1976). The discussion of natural kinds is perhaps better suited to the context of preparedness to learn—to acquire, to express—than to that of the language of natural categories (Ristau 1998).

The discussion in philosophy about natural kinds emerged to distinguish those representations structured in real events from those reflective of the cultural milieu. Of course, this distinction is nearly impossible to make. The natural kinds discourse was anchored to a view of the world that linked the problem of reference, innate quality space (Quine 1966, 1969), and the underlying ability to abduce the right hypothesis to a real property to be uncovered.

Other researchers have looked at the development of biological knowledge in relation to natural kinds. One investigator was concerned primarily with children's understanding of biological kinds (e.g., species; Keil 1992). In an experiment, children ages five to nine years were presented with various situations in which an animal appeared to transform into another animal by virtue of a costume or some other surface alteration or by injection with pills or vitamins (Keil 1992). For example, the subjects heard stories about a zebra in a horse costume, a lion with his mane shaved and black stripes painted on, baby chicks given a shot that made them grow black feathers and say "gobble, gobble," and so forth.

Among the youngest children in the study, 80 percent did not believe that putting on a costume was enough to truly transform an animal into another animal. Of this same group, 70 percent did not believe that a temporary surface change, such as painted stripes, was sufficient for transformation. In the more tricky situation of an injection or pill, in which subjects were to consider internal development as well as outward appearance, the youngest children were more likely than the oldest to accept it as a true transformation, but even 40 percent of the oldest agreed that it was an example of real change (Keil 1992).

Keil and his colleagues concluded that all children have beliefs about what constitutes membership in a biological group, and, for the most part, superficial changes do not affect those beliefs. Even five-year-olds have ideas about the mechanisms that underlie changes and about which of those mechanisms are biologically relevant to membership in a specific group. By age nine, some children are questioning the plausibility of the stories ("There are not really shots that can turn a chicken into a turkey"), considering the essence of biological kind ("When they were born, they were chickens. . . . They were chickens before they got their shots"), and even posing more advanced questions ("Does this shot have to do with the genetic structures of DNA to change the way it looked?"). Such comments demonstrate more sophisticated reasoning based on greater knowledge, and yet nearly half of this group were willing to accept an internal change combined with a change in appearance and behavior as example of real transformation (Keil 1992).

By the time children are in fourth grade, the relevance of natural kinds is clearly noted. Natural kinds are objects to get anchored to in forging a coherent world in which to act and adapt. One important attribute is the concept of an object. Apparently young children demonstrate knowledge of object permanency (Spelke et al. 1992; Spelke, Vishton, and Von Hofsten 1994) or object unity and substance (Millikan 1998). This knowledge is in part demonstrated in occlusion experiments in which the objects are partly seen or partly not. Infant gaze is the dependent measure. Of course, there are limits of studies that look only at fixation and habituation, but what else can one do with babies? They don't talk, and they don't do operants. Nonetheless, these kinds of results are quite interesting, and they suggest that theoretical orientation is there from the beginning.

Animate and Inanimate Objects

Young children are curious at an early age about the distinction between animate and inanimate objects (Carey and Gelman 1991; Keil 1996). It has been suggested that there is a fundamental category in our mind-brain for this distinction (e.g., Piaget 1929/1954; Premack 1990; Perani et al. 1999). Premack has further suggested that this fundamental category is important in coming to see others as having beliefs and desires, and is therefore essential in the development of social cognition.

Carey and her colleagues (e.g., Carey 1985; Johnson and Carey 1998) were concerned that Piaget's findings on children's tendency to attribute life to inanimate objects stemmed from a limited understanding of the word *alive* as adults use it (Carey 1985). She suggests that there are two sources of childhood animism. In some cases, children are, in fact, trying to answer the question posed to them: Is an object biologically alive? In other cases, however, children are trying to answer the wrong question; that is, based on an understanding that the opposite of alive is dead, they try to apply the distinction to an inanimate object, which, of course, is neither (Carey 1987; Carey and Gelman 1991).

Carey continues that at least one source of animism is an incomplete knowledge of biology (Carey 1987). She points out that never in Piaget's experiments or her own do children make judgments about life based on a single criteria or meaning; rather, like adults, they apply several criteria. Younger children (around four years old) tend to apply criteria that are irrelevant to biological life, such as whether an object exists or is useful. Later on, they begin to apply more salient criteria, such as movement and activity, anthropomorphic traits, and autonomous motion. By age ten, they have begun to apply such criteria as growth, reproduction, death, and composition. Carey sees this progression as the development of biological knowledge (Carey 1987) and, of course, it is also a sign of acculturation.

Carey also questioned whether Piaget's subjects were "trapped" into establishing a criterion for what is alive by virtue of the continuing series of questions. In an experiment designed to remove that possibility, investigators asked four-, seven-, and ten-year-olds about characteristics of certain objects, some familiar (dog), some unfamiliar (aardvark) (Carey 1985). Instead of asking whether the object was alive, they asked such questions as "Does it breathe?" "Does it need gasoline?" and "Does it grow?" The idea was to get the subjects thinking about objects in biological terms. They concluded that the children were relying on a biological framework, but their biological knowledge base was limited (Carey 1985).

Thus, Carey and her colleagues, with their realistic sensibilities, suggest that the concept of animism is real and constant throughout our lives. The storehouse of knowledge associated with this concept obviously changes as we get older, and in fact we learn to constrain the use of the concept. Carey concludes that animism also arises from a lack of biological knowledge (or restraint), and the acquisition of knowledge is partly responsible for the decline of animism in the first decade of life and throughout history.

Children at age five are in fact better able to generate correct answers to questions about living things than about dead things (Dolgin and Behrend 1984). When tested at age seven, they were equally good at answering questions about both. By five years of age, children, as we will see, are good at seeing others as not only as alive but with minds, who can deceive and be deceived and manipulated.

Animate Objects and Causation

The above discussion of animacy (biological kinds) is important, as I alluded, because it may bear on the notion of intentionality (Premack and Dasser 1991). However, to my mind it runs the risk of bringing back good old-fashioned dualism, not Cartesian dualism of substances, but conceptual dualism between mental and physical causation, which I believe is a mistake. On the other hand, there is no doubt a fundamental distinction between animate and inanimate objects (what a truism).

Premack (1995) has suggested that there is a basic distinction in our sense of causation that has relevance to the origins of human social understanding or competence. One sense is physical causation, such as billiard balls hitting each other. This experience is familiar. It is the stuff of physics; it is the paradigm of science. The other sense is traditional mental causation (e.g., Hempel 1965).

Animate objects can be described as both; they are self-propelled and moved by others. Premack interprets this duality from the point of view of the human, however; that is, he interprets events in terms of physical and mental causation. But one should be cautious here.

Perhaps we should think more simply about this cognitive adaptation to see others in terms of intentions. One suggestion is that we parse motion in terms of animate versus inanimate causation (e.g., Premack and Premack 1994; Caramazza and Shelton 1998). This cognitive capacity may be operative as a fundamental distinction in other species (Hauser and Carey 1998). Experiments with tamarin (*Saguinus*) monkeys have revealed that they are responsive to different kinds of causal events; that is, they have expectations of where an object will be depending on whether it is self-generated (a mouse) or not (Fruit Loops). A conception of self-propelled objects (i.e., animate) and of object motion (both animate and inanimate) would be useful categories. After all, part of the basis of theory of other mind is to infer self-causation (e.g., Premack and Premack 1994; Leslie 1994; as opposed to other sorts of causation (billiard balls hitting each other).

Suppose nature conferred to us distinct ways in which we interpret the world, one internally related to the ability to begin and stop behavior. This self-propelled sensibility is what James (1890) called the feeling of causation or what Whitehead (1927) called "casual efficacy" (see also Clark 1996/1997).

Infants and Intentionality

Infants, within the first few months of life, are clearly intentional in their own behavior (Frye and Moore 1991; Astington, Harris, and Olson 1988; Astington and Gopnik 1991). They may even recognize intentional communication, sharing gaze direction and contact; namely joint attention all of which participate in the development of social cognition (e.g., Brunner 1975a, b; Carpenter, Nagell, and Tomasello 1998). Researchers evaluated children at ages eight months, sixteen months, and two years in a simple "means-and-goal" task. A toy was placed on a blanket. The child was seated where she could not reach the toy but could reach the blanket. All of the children figured out that if they pulled on the blanket, they could get the toy. The goal is to reach the toy, and the means is the blanket; the consequence of pulling the blanket was to bring the toy within reach.

In one study, five-and-a-half-year-olds were shown short movies with two balls, one small and one large, and their interactions. The order of events was based on three scenarios, in which some interactions were intentional and others were not

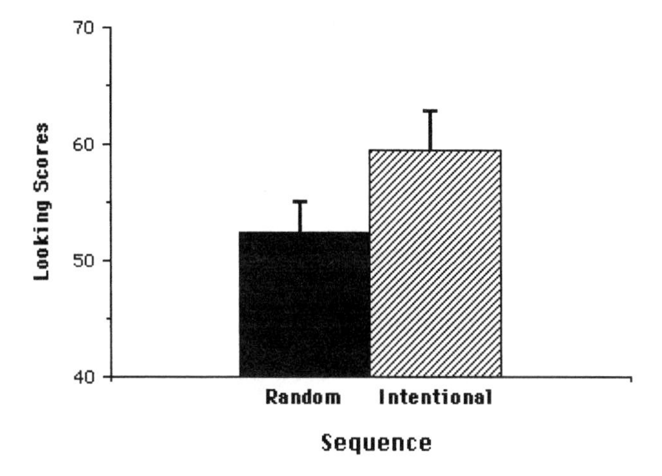

Figure 4.2
Looking scores for the control (random) and experimental (intentional) sequences in preschool-age children. Children looked longer at the experimental sequence after watching two previous control sequences than at the random sequence after two experimental habituation sequences (Dasser, Ulbaek, and Premack 1989).

(Dasser, Ulbaek, and Premack 1989), and complexity was controlled for. Children tended to look longer at the intentional event. (See figure 4.2.)

More recently, an experiment described at the end of last chapter Call and Tomasello (1998) demonstrated that in three-year-old children this ability to pick out the intentional (placing an object) from the accidental (placement of the object) is clearly evident. This finding is in contrast to the difficulty of demonstrating this ability in chimpanzees and orangutans and in younger children. (See figure 4.3.)

One claim, and I believe a misleading one, is that the attribution of desires precedes that of beliefs. The evidence for this hypothesis is that two-year-olds uttered more expressions with desire-based language than with belief-based language in a number of experimental paradigms (Wellman 1990; Bartsch and

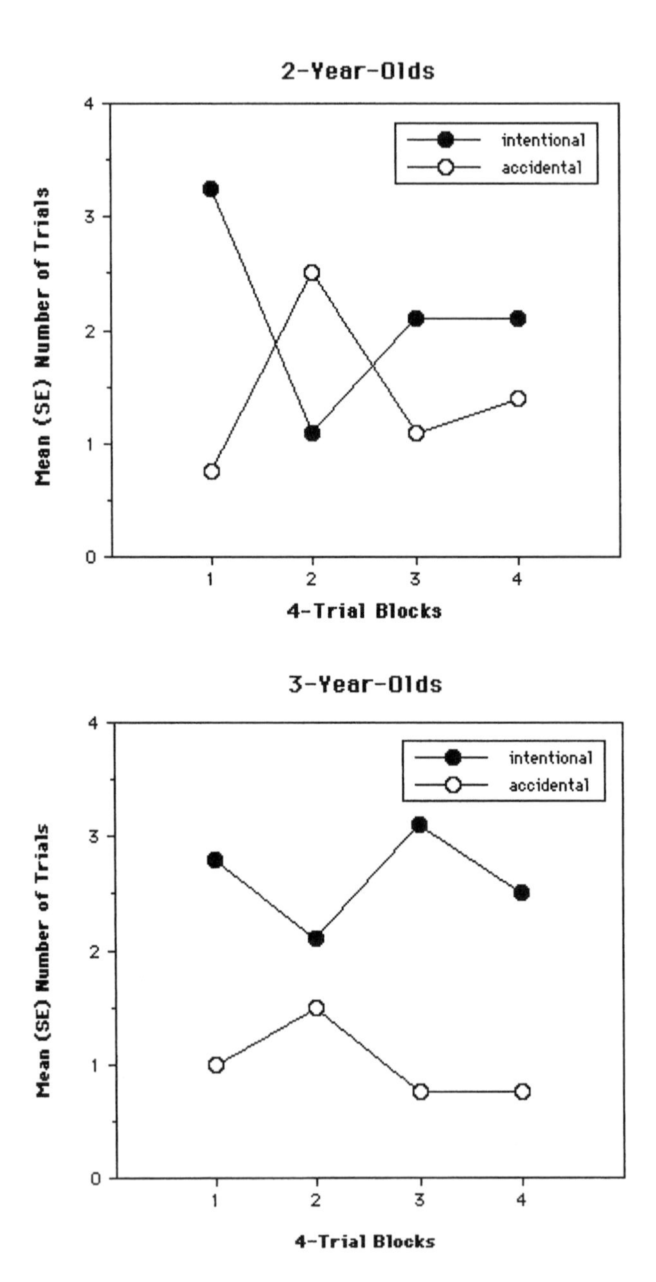

Figure 4.3
Mean number of trials in which two- and three-year-old children selected intentional or accidental actions as a function of four-trial blocks (adapted from Call and Tomasello 1998).

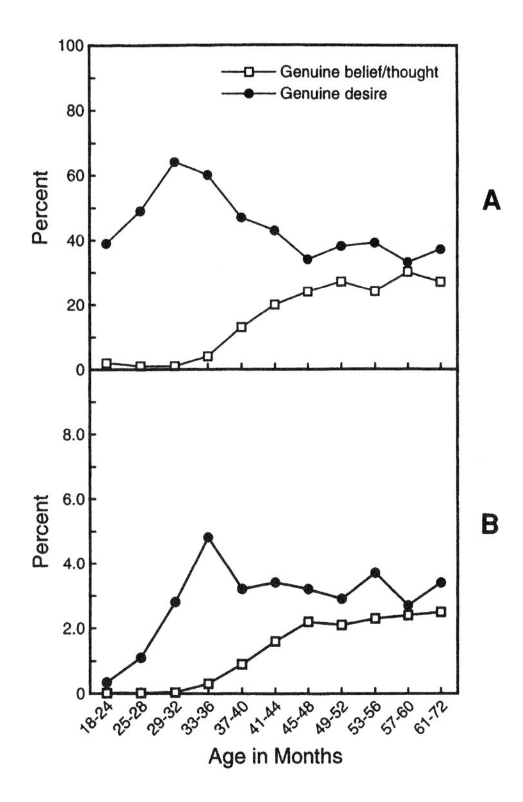

Figure 4.4
Psychological references to desires and beliefs, shown as a percent of all utterances using terms of desire and belief (A) or as a percent of total utterances (B) (Bartsch and Wellman 1995).

Wellman 1995; Wellman and Woolley 1990; Bartsch 1996). (See figure 4.4.)

By acting out with cardboard cut-out figures a series of stories about a character who wanted something, researchers tested whether two-year-olds were capable of simple desire reasoning (Wellman 1990). The character would look in one or two locations and find either what he wanted, nothing at all, or something else. Subjects were asked to make action judgments (e.g.,

"Will the character look in the other location?") and emotion judgments (e.g., "How does the character feel about finding . . . ?"). The subjects correctly predicted both actions and emotions based on the information provided to them about the character's desire.

To provide evidence for their theory that belief understanding follows desire in the developmental sequence of cognition, the same researchers went on to test the understanding of belief of the two-year-olds who had so successfully understood desires and actions. Using stories acted out by cardboard characters, the researchers asked the subjects questions about what the character would do based on the character's desires and beliefs. They instituted several mechanisms to separate the subjects beliefs and desires from those of the characters. As hypothesized, the children understood others' desires but were mostly unable to understand the beliefs of others.

It has been suggested that very young children can understand others' mental states in intentional but not representational terms. I am not sure what this means, considering that intentionality for us and not for the machines that we have so far invented presupposes the propositional attitudes, and the propositional attitudes are representations (Russell 1950/1980). Specifically, these young children understand that others want specific objects but not that others can believe certain propositions. Of course, none of this understanding may be conscious. Wellman and Bartsch (e.g., 1995) purport that a simple desire psychology allows children to understand a limited number of situations, but that increased understanding requires the ability to understand beliefs, essentially the next step up the ladder of social knowledge. I do not disagree with this argument.

These assumptions rest on perhaps a false distinction. Desires about apples already presuppose beliefs about apples; they are good to eat, and they are safe. It is not as if desires are not theory laden. Perhaps the difference between desires and beliefs is one of complexity, which is what I believe. This is not to say that beliefs and desires are not different. Surely they are. One such difference is that beliefs have truth value; desires often do not.

Again, Wellman and his colleagues suggest that desires attributed by a two-year-old are intentional but not representational. Why are representations reserved for knowledge states or beliefs? Surely representational abilities are a broader category? Everything in the brain at some point is representational. The question is one of degree. The fact that children younger than three do not reveal an ability to understand the propositional attitudes does not mean that their brains do not represent the world in terms of beliefs pregnant with desires and desires pregnant with beliefs. Neither is pure. Wellman and his colleagues make desires sound as if they are only itches to be scratched and as if they were not about anything. They make desires out to be only sensory and noncognitive, which they are not. The fact that a two-year-old can pass a reasoning test, described by Wellman with regard to desires, already implies that the desires are about something. The issue is whether it is easier to keep track of "desire beliefs" as opposed to detached distant beliefs. Both are propositional, however.

Appearance and Reality

The basis for distinguishing appearance from reality, imagination from truth, reveals something important that occurs at about three to four years of age in humans (Flavell 1986, 1988, 1999; Woolley and Wellman 1990; Wellman 1990). Again, the children express their cognitive competence earlier than Piaget thought. They in fact begin life parsing the world with cognitive categories (space, time). The distinction between desire and belief is not rigid, in just the same way perhaps the distinction is not as rigid between sense and thought. Beliefs pervade desire.

The child's sense of the mind goes through transition during development; she comes to see the mind as essentially active (Flavell 1999). The attribution of the active mind to a broad domain is curtailed with learning and development (Wellman 1990; Bartsch and Wellman 1995). They learn fairly readily to understand the role of imaginary mental representation and keep on distinguishing apperance from reality (Woolley and

Wellman 1990). Three- but not four-year-olds think imagination reflects reality. Children learn in this critical period the distinction between appearance in the mind and the facts in the world.

Early on, children explain behavior in multiple ways (Wellman 1990; Bartsch and Wellman 1995). Piaget thought young children were confused about behavior, but apparently they are less confused than he thought. They concurrently use both physical and psychological explanations for the same event, and they distinguish them readily.

A study evaluated how well four year olds could report past pretenses, images, perceptions, beliefs, desires, and intentions (Gopnik and Slaughter 1991). In a group of forty-three three- and four-year-olds, the researchers led the children toward a specific mental state (belief, surprise, etc.), then changed the mental state and asked the children to recall the first state. For example, in the belief task, children were shown a crayon box and asked what they thought was inside. The box was opened to reveal birthday candles. The children were then asked what they had thought was in the box before it was opened (i.e., their previous belief). The same type of question was asked after the children had pretended a stick was a spoon and then a magic wand (the pretend task), and so on for states of surprise, perception, desire, hunger, and intention.

Both age groups performed perfectly on the pretend task (i.e., they were able to pretend). In all the other tasks, four-year-olds did much better than three-year-olds. In the area of belief, nearly all the four-year-olds could correctly report their past false belief, but fewer than half of the three-year-olds could do so. Children of this age group are able to recall past mental states when those mental states do not require them to understand how representations are related to reality. The investigators note that this distinction underscores that the theory of mind begins to develop between the third and fourth year of life (Gopnik and Slaughter 1991).

Gopnik and her colleagues have also studied the development of a theory of perception in small children. By eighteen months, infants understand joint attention (Brunner 1975a,

1975b, 1983). For example, they realize that they can make someone else see what they see by pointing. More complex aspects of perception, however—such as the notion that the infant can see something that another person in the room might not (or vice versa)—come at a later age (Gopnik and Meltzoff 1997).

A study of children ranging in age from two to three years evaluated the relationship of age to development of perceptual understanding (see Gopnik and Meltzoff, 1997). The children were asked to mimic the experimenter in various tasks. One task was to hide or show an object so that the child could not see it, but the experimenter could, or vice versa.

Three-year-olds did quite well, thirty-month-olds performed at intermediate level, two-year-olds performed well at showing an object to the experimenter even when it meant that the child lost sight of the object. In the hiding tasks, two-year-olds failed to understand the perceptions of others: for example, they would hide the object from themselves by placing it on the side of the screen where the experimenter sat, or they would hide the object where neither of them could see it. Some children in the study, between the ages of two years and thirty months, would move the object several times, unsure of which placement would adequately hide the object, and some would walk to the other side of the screen to evaluate the placement from the perspective of the experimenter (Gopnik and Meltzoff 1997).

The understanding of visual perception that develops at around age three is more like that of adults and may be a sort of stepping stone to the understanding of belief that develops a little later (Gopnik and Meltzoff 1997). Researchers presented children with a picture that turned out to be a visual misrepresentation (the sensory equivalent of a false belief). They showed three-year-olds a picture of a cat that appeared black when viewed through a red filter but was actually green, and the children were generally capable of understanding the misrepresentation. These young children were better prepared to understand false-belief tests when these false beliefs were tied to perceptual misrepresentation (Gopnik and Meltzoff 1997). Other research has gone on to show that by teaching three-year-olds about the nature of perception, the children improve their understanding of false-belief tasks (see Gopnik and Meltzoff 1997).

In summary, children early on are capable of attributing beliefs and desires to others. Theory underlies perception and recognition of true and false beliefs, but the concept of theory is somewhat benign, not grandiose (Hanson 1958), and it is diverse (Heelan and Schulkin 1998). When it comes to other people, like other objects, we theorize about them. But in this limited sense (the concept of theory means more than what I have been elucidating; see Lewis 1972, 1983; and Botterill 1996), the child presupposes frameworks that are used to predict and understand events by postulating what goes on in the mind of others.

Children's Thoughts about Thinking

Anyone with young children will recognize the *why* question. They are endlessly asking why this and why that. I am constantly reminded of the limits of what I actually know. They are also philosophical (Matthews 1980).

In further investigations of what very young children think about mental processes, researchers have shown that three-year-olds do begin to grasp the concept of "thinking" but continue to have difficulty considering it in some contexts (Flavell, Green, and Flavell 1995; Shatz 1994). One researcher was blindfolded and held her hands over her ears while a second researcher asked the subjects questions about what the first researcher was capable of doing: Could she think about an object in the room? Could she think about another place? Could she kick her feet? The point was to assess whether the subjects could separate thinking from using the senses. The researchers concluded that by the time children are three, they understand that thinking about objects can take place with objects in and out of sight or sound (Flavell, Green, and Flavell 1995; Flavell 1999).

However, very young children do not appear to understand that people are always experiencing mental activity—the "stream of consciousness" James described (Flavell, Green, and Flavell 1995). Both three- and four-year-old subjects were asked to determine what a researcher was doing in a given situation. In one instance, the researcher was asked to try to solve a problem, and she then silently considered it. Only 13 percent of three-year-olds responded that she was thinking, compared with 88 percent of four-year-olds. In a similar test, the subjects were asked specifically if the researcher was "thinking about" the question, and the subjects fared much better. In the second situation, 79 percent of three-year-olds responded that she was thinking, compared with 94 percent of four-year-olds.

Subsequently, the subjects were asked if the researcher was variously "seeing," "talking about," or "touching" the object that was represented in the problem-solving question. Most (at least 94 percent) three-year-olds and all four-year-olds answered correctly, indicating that the children were clearly able to distinguish thinking from other activities that accompany thinking (Flavell, Green, and Flavell 1995).

Children's Choice of Frameworks: Behaviorist and Mentalist

As children start to attribute beliefs and desires to others, they enter the social arena. The world is larger than they, and they now take others into account. Moreover, by two years of age, children emit differential responses depending on whether they are referring to younger or older children (Flavell and Miller 1998). By three years of age, they choose a mentalistic explanation over a behavioral explanation in a number of contexts, which might reflect the time in which beliefs and desires come under the purview of the infant.

In other words, between ages three and four years, children also begin to take into consideration alternative representations of the same object (Gopnik and Astington 1988; Flavell 1999). Children three, four, and five years old were shown an object and then given further information that would change their perception of the object. Researchers tested their comprehension of representational change, false belief, and appearance versus reality. The children were shown a box of candy that turned out to contain pencils and a rock that turned out to be a sponge painted to look like a rock.

The researchers (Gopnik and Astington 1988) found that the three abilities required the children to consider alternatives and conflicting representations, and that all three were correlated at each age. All of the children performed worse on the question of representational change (What did you think before you knew the truth?) than on false belief (What would someone else think if they did not know the truth?) and on appearance versus reality (What does it look like? What is it really?). The investigators concluded that all three abilities involve alternative representations of reality and that this understanding develops in children between the ages of three and five years. Because all of the children performed much better at attributing false beliefs to others than at recognizing their own past false beliefs, the researchers hypothesize that children learn to understand changes in their own mental states by learning to understand such changes in others (Gopnik and Astington 1988). (See figure 4.5.)

Social knowledge begins to emerge, in part, at this point. Another team analyzed the effects of others on one's development of understanding by looking at siblings (Perner, Ruffman, and Leekam 1994; Jenkins and Astington 1996). Again using false belief as an indicator of the beginning of theory of mind development, researchers evaluated three and four year olds according to whether they had no siblings, an older sibling, a younger sibling, or a twin.

Three-year-olds with two or more siblings performed almost as well as four-year-olds (Perner, Ruffman, and Leekam 1994). These results have been extended and expanded to include young children (Jenkins and Astington 1996). The social interaction seems to facilitate the onset of the consideration of the beliefs and desires of others. (See figure 4.6.)

Moreover, it is not just the social contact that matters, but the degree to which one entertains a rich social life; the greater degree of fantasy, the better the performance on theory of mind measures (Taylor and Carlson 1997). Further, very young children can distinguish between psychologically caused actions

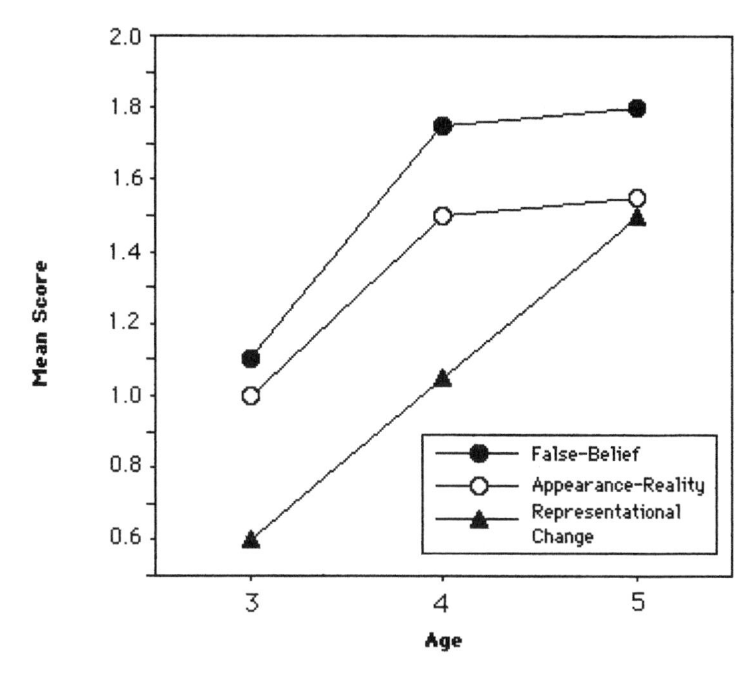

Figure 4.5
Mean scores on representational change, false-belief, and appearance-reality questions for three-, four-, and five-year-olds. Each child received a score of 0–2 for each type of question. Scores on all three types of questions improved with age. Four- and five-year-olds do well with the false belief questions (awareness that somebody else does not have the same knowledge as they themselves have) (Gopnik and Astington 1988).

and biologically or physically caused actions (Schult and Wellman 1997). That is, they realize that not all actions result from what one wants or desires—that uncontrollable outside forces affect our actions. Four-year-olds were told stories in which a character in a situation expressed a desire, and that desire was either fulfilled or thwarted as a result of a mistake (false belief) or by a biological or physical cause. For example, a character wants milk for his cereal but accidentally pours orange juice on the cereal (mistake/false belief: he thought the pitcher

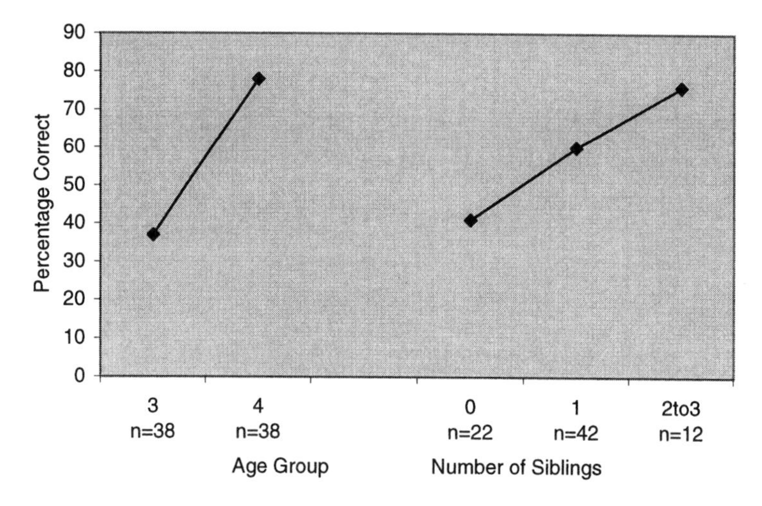

Figure 4.6
Percent of children giving correct answers to belief questions according to age and family size (Perner, Ruffman, and Leekam 1994).

of orange juice was a pitcher of milk). In another story, a character is hanging on a tree limb and says she would like to hang there forever, but she falls to the ground (biological cause: her muscles were fatigued). In another, a character standing on a stool wants to step off the stool and float in the air, but falls to the floor instead (physical cause: gravity does not allow people to float).

All of the children correctly identified the occurrences in the stories about intentions and mistakes as being the result of psychological causes. Sixty-nine percent of the children correctly identified the biological and physical causes in the corresponding stories. The researchers emphasize in their discussion that the four-year-olds were clearly not limited to psychological explanations for all situations. When the same test was given to three-year-olds, the results were difficult to interpret, but one finding stood out: the younger children were far more likely to change the facts of the story so that the character's desire matched the outcome of the story. The three-year-olds appeared to interpret actions as being the result of intentions.

However, when a group of three- and four-year-olds were told stories in which the characters wanted to perform a biologically or physically impossible actions, they were clearly aware of the constraints of nature, regardless of intention or desire (Schult and Wellman 1997). For example, they were asked whether a character could walk through a brick wall instead of going around it. Subjects were asked to give a simple yes or no answer, then prompted to give an explanation. The researchers found that almost all of the children in both age groups were capable of distinguishing voluntary, possible actions from desired but impossible actions. Some three-year-olds were even capable of providing sensible explanations based on physical and biological factors. (See figure 4.7.)

Knowledge and Imagination

In another study, three- and four-year-olds were tested on their comprehension of knowledge versus imagination (Woolley and Wellman 1993). Knowledge comes from direct perception of the world—seeing, touching, and so on—and children as young as three years begin to comprehend the link between the two. The authors questioned whether young children could distinguish between a mental image based on knowledge, such as a memory, and one not based on reality, such as a fantasy.

Most of the children were capable of making the distinction, including the youngest three-year-olds in the group. When the researchers evaluated the children's understanding of the relationship of knowledge or imagination to reality, the children three and a half years and older understood that knowledge reflects reality, whereas imagination does not. Children younger than three and a half years, however, often believed that both states represented reality.

The distinction between understanding mental states and conceiving of the mind as an independent actor or entity is apparent at six years of age. A study of children ages six to ten

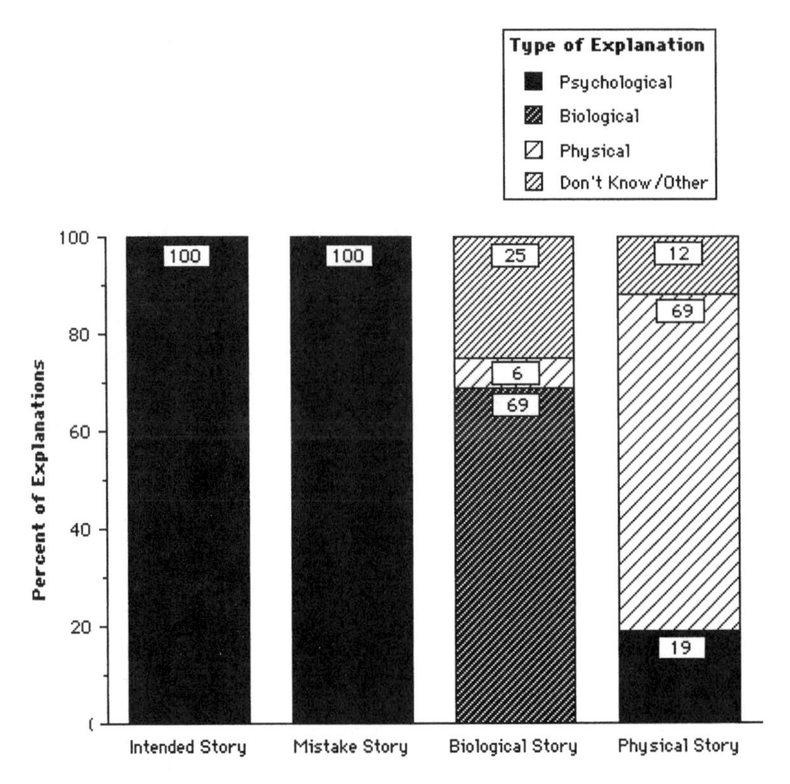

Figure 4.7
Children were asked to give reasons for story outcomes; reasons were psychological (getting the ketchup out of the fridge instead of chocolate sauce), physical (people can't fly), or biological (boy let go of bar because his hands were tired) (Schult and Wellman 1997).

years looked at this development by testing the subjects' comprehension of metaphors of the mind such as "My mind wandered" or "My mind was racing" (Wellman and Hickling 1994). The children were asked to interpret mind metaphors and, for comparison, mechanical metaphors such as "The car was sick" and natural metaphors such as "The wind howled."

Low scores were given to children who gave literal interpretations of a metaphor—for example, explaining "My mind was racing" as "He was running fast." High scores went to those who correctly interpreted phrases—for, example, explaining "My mind tricked me" as "You remembered something, but the memory wasn't right, even though you thought so." Overall, the children did well interpreting the car and wind metaphors, although the scores rose in correspondence with age, suggesting that most children understand personification as metaphor. In contrast, younger children scored very poorly in mind metaphors, whereas older children scored very high. The researchers conclude from these findings that children begin to understand and use metaphors of the mind in their mid-childhood years, and that this ability represents a developing understanding of the mind as an independent entity (Wellman and Hickling 1994).

Context and Knowledge

Perner and his colleagues claim that children interpret information based on a situational attitude—that is, a mental attitude toward a situation—not on internal or mental representations of situations: "I want *x*," or "He pretends *x*," or "She imagines *x*." The authors distinguish this type of understanding from understanding a mental representation of the situation. They believe that very young children fail to take into account another person's understanding (or representation) of a situation because they fail to make the connection between information and mental representation (Perner 1991).

To study this hypothesis, three- and four-year-olds were told two types of stories: one about knowledge and one about hunger. In the knowledge story, one boy is allowed to look into a box and see what it contains, but a second boy is not allowed to look. In the hunger story, one boy is allowed to eat dinner, but the second boy is not. After the researchers asked control questions to test whether the children understood what happened in the stories, they asked which boy knew what is in the box and why, and which boy was still hungry and why. Understanding clearly improved with age, but the real difference cropped up between the two types of stories. Children were much more likely to answer correctly and explain which boy was hungry than boy knew what was in the box. Apparently, it is much easier for young children to infer a state (hunger) than to impute knowledge or to infer what someone knows based on the information available.

Perner (Perner and Ogden 1988) says children understand a situation, not a mental representation of a situation, which distinguishes him from Leslie (1987; chapter 5), who believes that early pretense is an indication of understanding mental representation. Of course, none of this debate is about consciousness. It is an issue of the child's ideas about representations of objects—about context, action, and functional understanding.

Conclusion

Children early on are anchored to objects in their world; some objects to which we become anchored reflect our biological history and therefore are readily received from the environment. Moreover, early on we attribute intentions to others and are quite able to attribute beliefs and desires, learn about false beliefs, pretend, and deceive in predicting or explaining behavior (Taylor 1996).

Unlike in the animal literature, the findings about children show less ambiguity. Yet there are disagreements about the extent or necessity of attributing theory of mind to young children (e.g., Harris 1996, Harris et al. 1999). Even if simple desires with minimal belief structure appear before more elaborate beliefs, it does not necessarily mean that desire is closer to agency than belief. The child's understanding of agency includes both desires and beliefs; his knowledge and social practices are just quite limited. The extent to which this initial knowledge is confined to first- or third-person perspective making I think is unclear at this point (see Goldman 1993; Harris 1996; Carruthers 1996). I see no reason to return to a traditional view—namely, that we attribute to others only what we first discover in ourselves.

This distinction between desire and belief to be sure smacks of the empiricist distinction of desires unmediated by thoughts. But the idea that children are deeply concerned about the mental life of others seems quite palbable and real (Bartsch and Estes 1996). Desires are pregnant with beliefs about objects. Perhaps desires appear earlier because they are simpler. On this developmental account, desires appear first, similar to sensorimotor functions appearing before formal operations. But in my view, the sensorimotor explorations are highly cognitive or theory laden, as are the desires. Categories, information-processing systems in brain development, and even sensorimotor behavioral responses are always ideational (Dewey 1896; Jeannerod 1999).

A child's social knowledge is related to understanding the beliefs and desires of those around him. In the end, his cultural process and learning is intersubjective and must eventuate with collaborate bonds (e.g., Vygotsky 1962; Vygotsky and Luria 1993; Shantz 1975, 1983; Shatz 1994; Tomasello, Kruger, and Ratner 1993). Young children easily learn to collaborate with one another. The attribution of intentionality to others may be differentially expressed in the same neonate in different context and across cultures (Lillard 1997). Intentionality, inherently opaque, reflects degrees in development and appears throughout our lives as we attribute intentions to others and to ourselves as intentional animals.

It is the sense of other people and their wants—beliefs, habits, and so on—that pervades our social knowledge (Mead 1934). Our practices are social; we share basic frameworks of practice embodied in knowledge (Bourdieu 1972/1991). By necessity and definition, the entire social realm is intersubjective.

Little children learn, in part, about their social world by the attribution of desire and beliefs to others. Although some investigators in this field may overemphasize belief at the expense of desire, both are operative in the child's sense of her world. I do not believe, as I have indicated, that desires are independent of beliefs (see also Fodor 1992). However, both are fundamental in the intentional point of view.

Moreover, Lyons' (1995) assertion that nonlinguistic infants, although intentional, are without thought is patently false. In his very interesting book, he still makes it sound as if language is necessary for thought, but one important premise of the cognitive revolution is that thought is more than language. Nonetheless, his description of movement from a sensory and I would add cognitive level of intentionality to the linguistic level of intentionality to the propositional level of intentionality may in fact characterize the changes that occur in the development in our species. Moreover, there are many different kinds of cognitive systems, in which language is one among others.

Again, there is a danger in this newfound mentalism—namely, a backlash unless there are restraints on its application (see also Heyes and Dickinson 1990). Consider one example—pretend behavior. Children endlessly engage in pretend play, and in some experimental contexts they use theory of mind in problem solving (e.g., Cassidy 1998). I have watched my seven-year-old (eight by the time this book appears in print) and the range of her imaginary play behavior is fantastic. Dress-up play is something she does alone or with others. When she pretends, she may not always attribute or "be" anything more than behavior and may not take into account the experiences of or intentions of others. To narrow pretend play to intentional stances sounds extreme. But action programs within motor regions of the brain are replete with cognition (Berridge 1999; Graybiel 1995); to pretend, despite the fact that action programs are expressed (Harris 1989, 1996), does not mean that the action is not replete with intentionality (cf. Lillard 1998). The action may or may not be intentional. The constraint to inquire further whether the action is intentional or not is a safeguard against abuse. There is also the added burden of opaqueness as a feature of intentionality.

One further constraint to be considered seriously in the attribution of beliefs and desires or intention talk is to remember that the attribution regarding actions in the world—real, thought of, imagined—is not about consciousness. Moreover, to resist the fall into endless mentalism without constraint or to avoid the obvious backlash is to find constraints on these attributions; one constraint is to think of the world.

Even the issue of false belief, a critical marker in a theory of mind (i.e., nothing in the world corresponding to what is in my head), is not just a reflection of the mind, but the social world in which the child lives. Thus, the following principle: look to the practices, habits, and social structure of the child's world in understanding the attributions to others that the child makes about her beliefs and desires. There is no magic clarity of absolute perfection in such mental attributions. They are inherently fraught with imperfection and limitation. Such is the cost of being the sort of creature we are.

This social knowledge is not abstract. Moreover, there is a rich sense of social cognition without the theory of mind being evoked (e.g., Carpenter, Nagell, and Tomasello 1998). Nonetheless, parsing out social space through the attribution of beliefs and desires also makes references to the experiences of others. After all, we also believe that there is someone inside having an experience of one kind or another.

Perhaps the fact that autistic children do not collaborate reflects the fact that it is difficult for them to understand others and their beliefs and desires. We now turn to studies on autism.

Chapter 5
Autism

One popular hypothesis is that certain human beings may be inherently impaired in the ability to attribute beliefs and desires to others (autism). As Frith (1989) has put it, "In order to develop a coherent theory of mind one needs not only the ability to mentalize, but also experience. One needs experience with people who have different relationships to each other, and different personal interests" (p. 166). This impaired ability in autistic people may reflect alterations in specific regions of the brain that underlie social cognition of this sort (Baron-Cohen et al. 1999). The aberration of this ability reflects the breakdown of normal brain function, which Hughlings Jackson, the great nineteenth-century neurologist, referred to as the opposite of evolution (Jackson 1884).

In this chapter, I first describe autism—the evidence for a specific deficit in understanding other people's wants and desires (e.g., Leslie 1990; Baron-Cohen, Leslie, and Frith 1985)—then more global impairments (perhaps often not noted as much as they should be by theory of mind investigators), and finally which brain regions may underlie impairments in social cognition. The study of autistic children has revealed indeed great impairments in paying attention to the experiences of other people. The sensory impairments (the body as a vehicle for knowing) often noted in autistic children and adults may reflect defects in social information–processing system in the brain. Moreover, I take issue with some of the talk about metarepresentation and introspection.

Autism

It has long been recognized that autistic children are socially withdrawn (Kanner 1943; Asperger 1944; Frith 1991/1997; Baron-Cohen et al. 1995, 1999), which was once thought to be due to environmental factors (Tinbergen 1974), although it is now noted to result from considerable genetic determinants of behavior. Also, more males than females are autistic (Santangelo and Folstein 1999), and we all know that females have greater social intelligence.

Autism effects about four to fifteen children per ten thousand. It cuts across class and habitat. Early on, it was described by Kanner (1943) in the context of withdrawal, loss of contact (eye), and lack of responsiveness to surrounding social facts. Not making contact with others is a key feature of the condition. The "other" barely exists. (See figure 5.1.)

Attachment behaviors are compromised in autistic children (Sigman and Ungerer 1984; Sigman et al. 1986). Attachment is at the heart of our social relationships—at least the basic ones that get us started and provide stability and security. There are also varying degrees of autism (Siegel 1996; Volkmar 1998). The *Diagnostic and Statistical Manual of Mental Disorders* (DSM) criteria for autism are depicted in table 5.1. Autism can also be indicated by abnormal or impaired development prior to age three, which is manifested by delays or abnormal functioning in at least one of the following areas: (1) social interaction, (2) language as used in social communication, or (3) symbolic or imaginative play. It is clear that autism includes a range of behavioral abnormalities that reflect the degradation of normal human contact.

Autistic patients also have blunted affect (Hobson 1984, 1993) or dampened emotional responses, and their sensorimotor or bodily responses are compromised. Many autistic individuals also act as if sensory stimuli are painful, distracting, or aversive. For them, the body is not a vehicle with which to engage the world (Hobson 1993), which is reflected by their sensorimotor impairments. These sensorimotor impairments could contribute

Figure 5.1
Characteristic traits of autism (Frith 1993).

Table 5.1
Diagnostic and Statistical Manual of Mental Disorders, Fourth Edition (DSM-IV), criteria for autistic disorder

A. Qualitative impairments in reciprocal social interaction

1. Marked impairment in the use of multiple nonverbal behaviors such as eye-to-eye gaze, facial expression, body posture, and gestures to regulate social interaction

2. Failure to develop peer relationships appropriate to developmental level

3. Lack of spontaneous seeking to share enjoyment, interests, or achievements with others

4. Lack of socioemotional reciprocity

B. Qualitative impairments in communication

1. A delay in or total lack of the development of spoken language (not accompanied by an attempt to compensate through alternative modes of communication such as gesture or mime)

2. Marked impairment in the ability to initiate or to sustain a conversation with others despite adequate speech

3. Stereotyped and repetitive use of language or idiosyncratic language

4. Lack of varied spontaneous make-believe play or social imitative play appropriate to developmental level

C. Restricted, repetitive, and stereotyped patterns of behavior, interests, or activity

1. Encompassing preoccupation with one or more stereotyped and restricted patterns of interest, abnormal either in intensity or focus

2. An apparently compulsive adherence to specific nonfunctional routines or rituals

3. Stereotyped and repetitive motor mannerisms (e.g., hand or finger flapping or twisting in complex whole body movements)

4. Persistent preoccupation with parts of objects

To be diagnosed with autistic disorder at least one sign (each) from parts A, B, and C must be present, plus at least six signs overall.

to the withdrawal from social events and therefore from the life of others.

Autism and Knowledge of Others' Experiences

One claim is that autism is the loss of a theory of mind, a critical feature of social cognition (Baron-Cohen, Leslie, and Frith 1985; Leslie 1990). Autistic children are more impaired than this claim suggests (Baron-Cohen et al. 1995, 1999; Tager-Flusberg 1999); as I indicated above, they do not liked to be touched, and they are withdrawn in general. Moreover, it has been hypothesized that a special module or mechanism for understanding the beliefs and desires of others exists in the brain (Baron-Cohen et al. 1999; Leslie 1990), a proposal that is controversial (e.g., Harris 1989; Perner 1993) but interesting.

Let's look at several key experiments. To demonstrate the early expression of the theory of mind, one experiment uses a puppet show (Baron-Cohen 1995; Leslie 1987, 1990, 1994). One puppet ("Sally") has a marble. She places the marble in a basket and leaves the room. Another puppet ("Ann") takes the marble out of the basket and places it in a box in the same room. Sally comes back. The subjects are asked where Sally will look to find the marble.

Generally, autistic individuals have great difficulty with this question. They will answer that Sally will look for the marble in the box. They are unable to comprehend that, although they saw Ann move the marble from the basket to the box, Sally does not know that the marble has been moved. That is, they are unable to attribute a false belief to Sally. By age four, normal children can attribute false beliefs and distinguish their own knowledge from that of others (Baron-Cohen 1995; Leslie 1990).

Another paradigm, pretend play, has suggested that autistic kids reveal defects in theory of mind reasoning (Leslie 1990). In this paradigm, one has to imagine or pretend to be someone else, or "act as if." It is claimed that the ability to pretend is linked to the ability to pretend to be someone else, to take another person into account—which is to say, to have social

Mother's behaviour:

talking to a banana!

Infer mental state:

mother PRETENDS
(of) the banana (that)
"it is a telephone"

Figure 5.2
A pretend scenario (after Leslie 1987, 1994).

knowledge. One is forced to consider others' wants, beliefs, and desires.

The model of pretend play is depicted in figure 5.2. A person (mother) pretends that a banana is a telephone. She is talking to the banana. As the researcher describes it, the relationship of one object (mother) to another (banana) is "minimally interesting, [but] the real significance of her behavior emerges only when mother is described as an agent in relation to information" (Leslie 1990). Mother is pretending, and children as young as two years old are capable of understanding that mother is pretending. The ability to interpret mother's intentions in relationship to the physical facts of the situation is described as *metarepresentation* (Leslie 1990; Baron-Cohen 1995). Autistic children do not comprehend that mother is pretending that the banana is a telephone.

Consider another example. Autistic children and normal three- and four-year-old children were given false-belief tasks (such as the doll scenario in which Anne moves the marble while Sally is out of the room). They were also shown similar false photographs (for example, Sally takes a Polaroid picture of the marble in the basket before she leaves the room, Baron-Cohen, Leslie, and Frith 1985). The normal children found the two tasks about equally difficult; the autistic children could not accurately respond to the false-belief task but passed the false photograph task.

The authors (Leslie and Roth 1993) conclude that autistic child can represent objects, but they do not understand that pictures, although representations, are not agents. But this depiction renders pictures as merely sensory, and once we move from the view that pictures are sensory, the distinction becomes dubious.[9] The autistic child viewing pictures of objects might have some beliefs and indeed does.

Some researchers believe that autistic children have an impaired capacity to metarepresent social events, which is in part why they are incapable of imagining the beliefs of other people and of being able to pretend (cf. Leslie 1987, 1990; Leslie and Roth 1993/1999; Perner 1993/1999). But the utility of the concept of metarepresentation in this context is dubious because it means stacking on representations and runs the risk of borderline infinite regress—namely, endless represenations. Moreover, again I get the sense that consciousness is slipping back in. What stands out in studies of autistic children is that they are more impaired in the ability to imagine the beliefs and desires of others. Some researchers have disputed the idea that autistic children lack a theory of mind simply because they fail at metarepresentation (a confusing and misleading term) tasks (Leekam and Perner 1991; Perner 1993/1999). Of course, complicated cognitive functions are indeed impaired in autistic children, such as the ability to represent complicated social relationships.

Autism as a Fundamental Impairment in Social Knowledge

Another example of autistic children's impaired social knowledge can be seen in their understanding of attitudes conveyed by others in conversation. By five years of age, most children have a vast amount of knowledge of the beliefs of others and how those beliefs are reflected in conversation (Flavell 1999). Autistic children, however, absorb much less from conversation than normal children (Roth and Leslie 1991). The pragmatic use of language in social discourse is impaired in autistic children (e.g., Tager-Flusberg 1981, 1993/1999).

An experiment designed to compare the relative abilities of normal three-year-olds, normal five-year-olds, and autistic adolescents to comprehend others' beliefs through conversation demonstrated this point (Roth and Leslie 1991). The subjects watched an exchange between two dolls, Yosi and Rina, who are playing ball together. Yosi has some chocolates, which he places on the ground. The ball goes behind a house, and Yosi goes to get the ball. Rina hides the chocolates in a box. Yosi returns and asks what happened to his chocolates. Rina says the dog took them. When Yosi asks where they are, Rina says they are in the doghouse.

The subjects were asked where Yosi thinks the chocolates are, where Rina thinks the chocolates are, and where they (the subjects) think the chocolates are. Most five-year-olds could answer all the questions correctly, meaning they were able to understand that Rina lied to Yosi, that Yosi believed what Rina stated, and that Rina was, in fact, aware of the deception of her statement. Most three-year-olds apparently attributed to Rina not intentional deception but mistaken belief. Most of the autistic subjects, on the other hand, answered all the questions based on the reality of the situation—that the chocolates were in the box. They were unable to grasp the intentions and beliefs of the dolls in the play from their conversation.

Another study compared autistic children with IQ scores in the normal or borderline range of intelligence with children of equal mental age who had specific language impairments

(Leslie and Frith 1989). These two groups were selected to ensure that differences in understanding the experimental questions were not based on language comprehension problems. The researchers presented a scenario much like the doll scene described above, except that the researchers themselves were the actors. They wanted to test autistic children's comprehension of limited knowledge (i.e., the fact that someone else does not have full knowledge of the situation even when the subject does) and of false belief (i.e., given limited knowledge of the situation, someone else may wrongly believe something to be true).

In the scenario, one person leaves the room and another moves an object from one hiding place to another in full view of the subject. To test their understanding of limited knowledge, the subjects were asked whether the first person knows the object has been moved and where that person would look for the object. To test their understanding of false beliefs, subjects were asked where the object really is and where the first person thinks it is. Confirming their hypothesis, the researchers found that the autistic children still had greater difficulty understanding the concepts of both limited knowledge and false beliefs than did the language-impaired children. This finding has been construed as proof that the metarepresentational capacity of autistic children specifically is limited.

Another experiment to evaluate the limitations of autistic children's theory of mind is a series of picture stories. Each story depicts a situation, cause of action, and effect (see figure 5.3) (Leslie and Frith 1990; Baron-Cohen, Leslie, and Frith 1985). The first story shows a boy kicking a rock, which is an example of a person interacting with an object and is termed a causal-mechanical situation. The second story shows a girl taking an ice cream cone away from a boy and is termed a social-behavior situation. The third shows a boy taking away a girl's teddy bear while her back is turned and is termed an intentional situation—one that can best be understood by inferring the mental state of the girl when she finds her teddy bear missing.

Compared with controls (which included children with Down's syndrome and normal preschoolers, all of whom had

Figure 5.3
Illustration of the three story types: (a) causal-mechanical (person-object subtype); (b) social-behavioral (person-person subtypes), and (c) mental state (after Baron-Cohen et al. 1985; Leslie 1990).

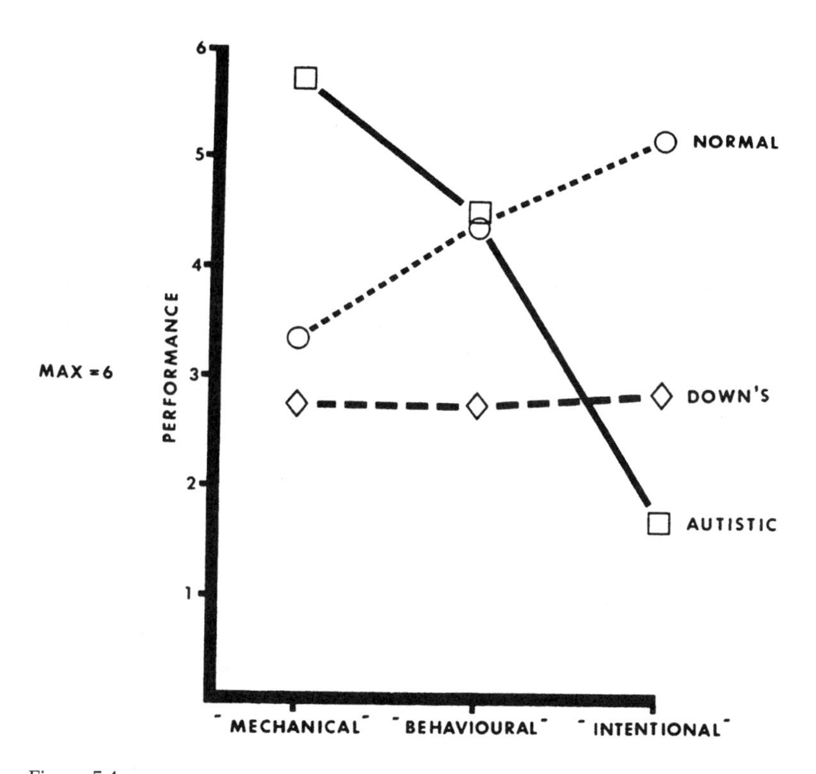

Figure 5.4
Results of picture-story sequencing task (after Baron-Cohen, Leslie, and Frith 1985; Leslie 1990).

lower mental ages than the subjects), the autistic children clearly understood the causal-mechanical situation and displayed a fair understanding of the social-behavioral situation. As was expected, however, they had great difficulty understanding the intentional situation. The researchers noted that the autistic children were able to comprehend situations of social interaction that involved a cause and effect, but the children specifically had trouble with a situation in which it was necessary to understand what someone else knows or expects. Figure 5.4 depicts the performance of all three groups of children in comprehending the three situations.

More on True and False Beliefs

Many studies have also looked at the ability to distinguish true and false beliefs. One study, for example, assessed true belief and false beliefs in normal three-year-olds and autistic kids. Compared with three-year-olds, autistic children are impaired with regard to false belief (Perner, Leekam, and Wimmer 1987; Leslie 1990).

Perner and his colleagues have argued that an understanding of false beliefs emerge at about four years of age. False beliefs give rise to the recognition that the events in the speaker's world may not have a reference and that the world is now being reshaped away from that of strict reference.

For example, autistic children were shown a box of a familiar brand of candy and asked what it contained (Perner, Leekam and Wimmer 1987). The children presumed that the candy was in the box but were then shown that the box contained only a pencil. They were asked what the next child to come to the room would say when shown the same box and asked what it contained. Based on their own knowledge, the children thought that another child in the same situation would answer that the box contained a pencil. Autistic children have difficulty understanding that others don't know what they know. And, of course, people to varying degrees make this assumption. Autistic individuals are just the extreme. The range of variation and of ability is a hallmark of a biological phenomenon.

As many as 35 percent of autistic children do pass tests of false-belief understanding (Baron-Cohen 1995 1996). However, this percentage does not imply that these children have a normal theory of mind development (Baron-Cohen 1995 1996). In further tests, such children are successful with first-order beliefs—for example, they can discern where Sally thinks the marble is—but not with second-order beliefs; that is, they cannot discern what Ann thinks that Sally thinks, something most normal six or seven year olds can comprehend (Baron-Cohen 1995 1996).

A larger class of events should also be considered in the context of false belief: deception and lying. One consequence of deception is mistrust; another is a greater ability to discern truth from falseness. The recognition of intentions is linked to this ability. Children's ability to discern false belief increases in the fourth year of life. Young children also have no difficulty deceiving regularly by age three or four years (Wellman 1990).

Autistic patients, on the other hand, are impaired in the ability to deceive, according to some researchers. A study of deception compared autistic children with normal three-year-olds and children with mental disabilities (Baron-Cohen 1995). The subjects were asked to hide a penny in their hands. The autistic children were able to keep the penny hidden but often neglected other aspects of the deception, such as closing the fist of the empty hand. The other subjects were much more likely to take other aspects into account—that is, to consider what the guesser might think about where the penny was hidden. In another study of autistic, mentally retarded, and normal children given a one-on-one task of deception demonstrated that autistic children were more impaired than the other two groups in their ability to perform this task (Frith 1997).

A theory of mind may require the sensibility that objects are self-propelled (Premack 1990; Leslie 1990; Baron-Cohen 1995). Other cognitive factors include the ability to make eye contact or follow the eye gaze of another and to generate triadic representations (e.g., Lee et al. 1998; Frith and Frith 1999), and, finally, knowledge of mental states used to predict behavior (Baron-Cohen 1996). Of course, I would add something else: understanding something of the experiences of others.

Shared Attention and Autism

As discussed in earlier chapters, the ability to look at an object and know that another person is doing the same is termed *joint attention*. This ability appears to be impaired in autistic children. For example, autistic children do not demonstrate "gaze monitoring"—that is, looking back and forth from an object to the eyes of another person to determine whether the other person is also looking at the same thing. Autistic children will point or gesture to an object that they want (the "protoimperative gesture"), but they have difficulty pointing or gesturing to an object for the sake of drawing another person's attention to it (Baron-Cohen 1989a, b, 1995).

The point is that, in fact, autistic children seem to have little interest in what others around them see, hear, or experience (Baron-Cohen 1995). The only time they will point out an object, present an object, or lead someone to an object is when they want assistance getting or using the object (Baron-Cohen 1995). It has been postulated that the fact that autistic children often speak either too loudly, too softly, or without intonation shows that they lack the concept of an interested listener (Baron-Cohen 1995). They do not seem to respond normally to this social feedback. For example, in one study of spontaneous play, only one autistic child (10 percent) and two developmentally delayed children (22 percent) appeared to pretend play, compared with eleven normal children (63 percent).

In a joint attention task, the subject and his or her parent watched as a researcher worked the switch to an automated toy. (The toy had a cord that attached it to the control box so that it was clear that the researcher's actions were controlling the toy.) Almost all of the normal and delayed children looked from the toy to the parent, whereas almost none of the autistic children did. In contrast, approximately half of the subjects in each group looked from the toy to the researcher. Among the autistic children, it was clear that they were interested in the connection between the movement of the toy and the researcher's control of it, as were the other children, but not in sharing the sight with the parent.

Finally, subjects were asked to imitate the researcher in a play task. Not surprisingly, most of the normal children (80 percent) successfully imitated the researcher, half (53 percent) of the developmentally delayed children did, but less than a quarter (22 percent) of the autistic children did (Baron-Cohen, Leslie, and Frith 1985; Baron-Cohen 1995 1996).

Many researchers have concluded from these experiments and others that the developmental differences—namely, those of a social nature—between autistic children and children with other mental disabilities are present very early in life. A screening tool called the Checklist for Autism in Toddlers (CHAT) allows researchers to detect those children at risk for autism as early as eighteen months of age. Researchers compared a group of such children with a normal control group and a group of developmentally delayed children to determine if some of the signs of a lack of theory of mind mechanism could be seen so early in autistic children (Charman et al. 1997). They measured empathic response, spontaneous play, inclination to pretend play, joint attention, and imitation ability.

In the empathy task, the subject and the researcher played together with a toy hammer. The researcher pretended to hit his thumb with the hammer, crying out and making faces as if in pain, and, after a few seconds, showing the child that he was not really hurt. All of the normal and developmentally delayed children looked at the researcher's face when he pretended to hit his thumb, but only 40 percent of the autistic children did. Most of the normal children and nearly half of the delayed children had facial expressions of concern, whereas none of the autistic children did.

A further test of theory of mind in autistic or Asperger syndrome (a syndrome with some of the characteristics of mild autism; see Frith 1991/1997; Wing 1991/1997) is inferring mental states by information from photographs of eyes (Baron-Cohen et al. 1997). Normal children can do this rather easily, as can adults, but both autistic and Asperger syndrome subjects with normal IQ have great difficulty with this sort of task. Control subjects and Tourette's syndrome subjects did not. Subjects were demographically matched. Again, it is the idea that the eyes are or rather can be revealing about the mind, and again they can or cannot. (See figure 5.5.) On the other hand, the eyes are a useful marker of the person's sensibilities, and we use this marker rather easily and readily. Thus, children with autism do not show common forms of joint attention (Baron-Cohen et al. 1997) and are impaired in imitation.

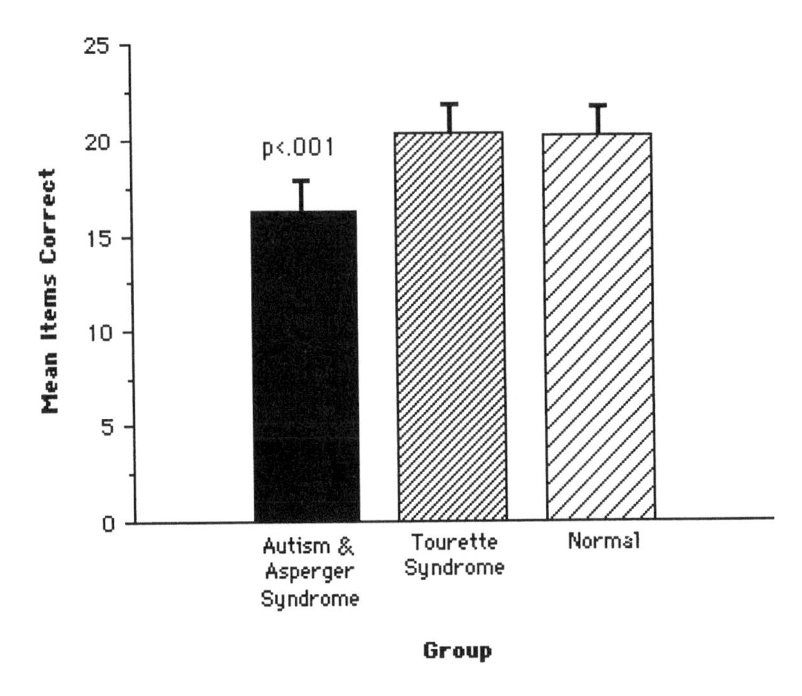

Figure 5.5
Performance on the eyes task by patients with Asperger syndrome and Tourette's syndrome and by controls. Patients with autism and Asperger syndrome do worse on a task where they are asked to identify which words best describe the thoughts and feelings of people in photographs (Baron-Cohen et al. 1997).

Theory of Mind and the Brain

A growing literature suggests that some key neural structures may underlie the attribution of intention to others and social knowledge (Frith and Frith 1999; Adolphs 1999). There is some evidence to suggest (Baron-Cohen et al. 1994; Happe et al. 1996; Happe, Winner, and Bronell 1998; Fletcher et al. 1995; Goel et al. 1995; Sabbagh and Taylor 2000) frontal cortex involvement with the attribution of beliefs and desires to other people's experiences (see also chapter 6). (See figure 5.6.)

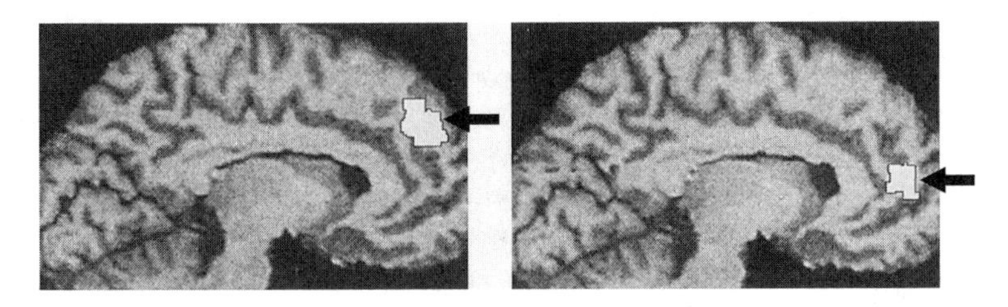

Figure 5.6
Brain scans show differences in activity between normal and autistic people. In normal persons reading a story that requires inferring the mental state of others, the left medial prefrontal cortex of the brain was active (left). In persons with Asperger syndrome performing the same task, an adjacent lower area was active instead (right). The left medial prefrontal cortex may be a key component in theory of mind capability (Frith 1993).

Using photon emission computerized tomography (SPECT) to measure blood flow as an index of neural activation, studies show an increased activation of the frontal orbital field when subjects are asked to think about mental as opposed to physical words (e.g., Baron-Cohen et al. 1994). Through headphones, subjects listened to words that depicted the functions of the mind (issues about beliefs, desires, emotions, pretend situations, etc.) interspersed with random words. In a comparison test, the same subjects were told to attend to words about bodily function (walking, blood, food intake). The results suggested that there was increased activation (greater blood flow) in the orbital frontal cortex when subjects were attending to terms about beliefs and desires than when they were attending to bodily considerations per se.

In another study that used PET, subjects activated the left medial prefrontal cortex while they were reading stories requiring inferences about mental states. This was not the case when they were reading stories strictly about physical functions (Fletcher et al. 1995). Other regions that were activated to a greater extent by the theory of mind stories than by physical stories were the bilateral temporal poles—the left superior temporal gyrus and the posterior cingulate cortex. (See figure 5.7.)

In a further study using fMRI, subjects were asked to perform both verbal and nonverbal tasks in which one context would require a theory of mind attribution and another would not (Gallagher et al. 2000). Subjects were shown cartoons in both contexts. For the context in which the attribution of theory of mind might be required, there was greater activation of the medial frontal gyrus and anterior cingulate regions of the brain.

In a PET study of Asperger syndrome patients, subjects (controls and Asperger patients) were matched for intellectual functions. The results revealed that the left medial prefrontal cortex is much less active in the Asperger subjects than in normal control subjects when they are forced to make inferences in obvious stories about mental functions (Happe et al. 1996; Frith 1993). (See figure 5.6.)

Finally, another study using fMRI to assess social intelligence in normal and autistic brains have revealed several interesting facts (Baron-Cohen et al. 1999). First, the authors found that three regions of the brain that had earlier been thought to underlie social cognition were active in theory of mind tests (e.g., regions of the frontal and temporal cortex and amygdala; Brothers 1990). The subjects were asked to consider mental states in photographs of eyes. The autistic groups and controls were matched for IQ and left- or right-handedness.

The brain regions that were activated in controls were the orbito-frontal cortex, superior temporal gyrus, and the amygdala (Baron-Cohen et al. 1999). Autistic subjects showed less activation in both cortical sites and showed no activation in the amygdala when compared to normal subjects. In some detail, in the next chapter, I describe the role of the amygdala, which is importantly involved in emotional judgment (Rosen and Schulkin 1998).

In general, the brains of autistic people are different from those of normal people in a wide array of regions. In general in autistic people, there is less activation in the prefrontal cortex for a number of tasks, as revealed by fMRI (Ring et al. 1999). A wide net of neural sites have been linked to autism in addition to other cognitive disorders (Happe and Frith 1996; Waterhouse,

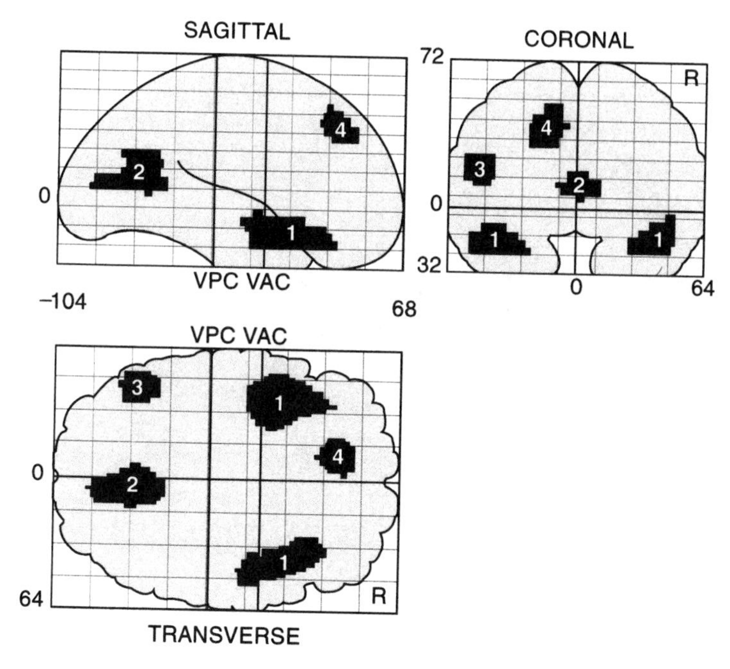

Figure 5.7
Statistical parametric maps (SPMs) of the comparison of the theory of mind task with the unlinked task. (1) The temporal poles bilaterally, (2) the posterior cingulate cortex, (3) the left superior temporal gyrus, and (4) the left medial frontal gyrus (Fletcher et al. 1995).

Fein, and Modahl 1996). Regions of the frontal, temporal, and parietal lobe and the anterior cingulate gyrus have been implicated in autism (Mountz et al. 1995; Courchesne et al. 1994). Disruption of normal expression of cranial nerves, cerebellum (Harris et al. 1999), and reduced size of the corpus callosum (Piven et al. 1990) are features of autism, as well as delayed maturation of the frontal cortex (Zilbovicius et al. 1995), alterations of hippocampal and amygdala function (Bauman and Kemper 1994, 1997; Baron-Cohen et al. 1999), and basal ganglia (Damasio and Maurer 1978) or cortical connectivity (Happe and Frith 1996). Finally, there is some evidence that

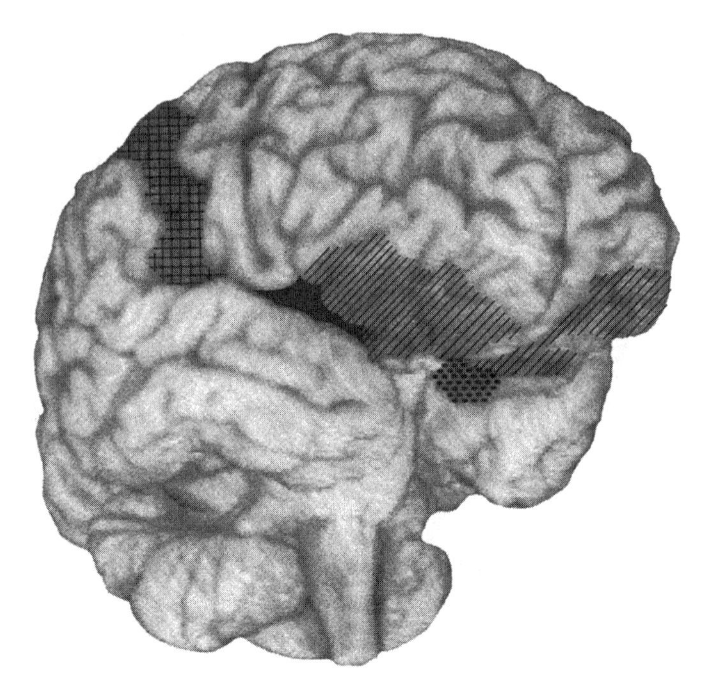

Figure 5.8
Some of the neuroanatomical structures involved in social cognition. Diagonal lines are within the ventromedial prefrontal cortex, black area is within the insula cortex, dotted area is the amygdala, cross-hatching is within the somatosensory cortex (courtesy of Ralph Adolphs).

the right hemisphere may be linked to this social-parsing ability, with which the beliefs and desires of others can be discerned (Winner et al. 1998; Happe et al. 1999), and to general executive function in autistic children (Happe and Frith 1996).

Conclusion

Humphrey (1976) envisioned social knowledge and the introspective organ as tied to a theory of mind. As I have indicated, I think this way of speaking about an "introspective organ" is misleading and mistaken.

Pretend play is one paradigm that can be construed in terms that are less about mental states and more about actions (Harris 1996). Of course, action is replete with mental states, so this distinction is moot (chapter 6). The fact that pretend play recruits action patterns of goal-directed activities does not mean that the action is not laden with thought. The question of whether we need to attribute an elaborate mental event is, however, a prophylactic against the ease of attributing something special going on inside the subject—the fallacy of the looking in on the mind's activities.

Social parsing (e.g., the attribution of beliefs and desires to others) is a cognitive/behavioral adaptation. The ease with which we attribute intentions may facilitate the learning about the social structure we are trying to adapt to and to understand. The social structure is already there. We step into social structures that are established and to which we accommodate and adapt. The social structure and its meaning are the furniture of the world and not in isolation in the head.

Therefore, the results of the studies reported above are consistent with the view that autistic children are minibehaviorists in part because they do poorly separating physical from mental tests; they are also poor in considering their own mental states when compared to retarded and normal controls (Baron-Cohen 1995; Leslie 1990). Moreover, autistic children have particular trouble understanding when emotions are linked to false beliefs

(Baron-Cohen 1995) or distinguishing jokes from lies (Winner et al. 1998). They are limited in playing games, such as penny hiding, in which deception figures prominently.

Interestingly, in imperative communication tests of problem-solving abilities involving others, autistic children sadly did not treat the others as subjects, but rather viewed them as objects (Philips, Baron-Cohen, and Rutter 1992). In other words, they did not appreciate that others had experiences—had "insides." But, then, autistic children also have trouble with attentional tasks (e.g., Pascualavaca et al. 1998), just as some disabled people display deficits in understanding others (e.g., Peterson and Siegal 1999).

Does this mean that the autistic person is blank with regard to the beliefs and desires and experiences of others? The answer is no (see Harris 1989, 1996). They are just severely limited in these domains. The limited response to other people's emotions, the blunted affect, and the enhanced sensory responses of autistic people can be painful and disorganizing in some contexts, which is not inconsistent with a limited knowledge of others. The extent to which there is both first- and third-person deficits is not clear in the case of autism, and, of course, the two are intimately joined.

Does autism really provide a context in which social knowledge is separated from other sorts or in which the mechanisms of this sort of cognition are distinct from others? In addition, the extent to which social knowledge is developed as a function separate from other cognitive functions is not clear. It seems reasonable to me to posit a cognitive function or a set of them devoted to social knowledge and to state that social practices are vital to and underlie social knowledge (Sabini 1992). Obviously, cognitive resources are utilized in this process (Waterhouse and Fein 1997). Remember, the person is unconsciously adapting or using cognitive resources. Also remember, and sadly, that the autistic child does not benefit by understanding or acknowledging the experiences of others.

Finally, with regard to the neural circuitry, certainly the prefrontal cortex and the amygdala are involved in a wide array of functions that serve us in the organization of action and that underlie a specific ability to interpret others' beliefs and desires (Baron et al. 1993/1999). In addition, there are findings of aberrations in sensorimotor function and perhaps in basal ganglia function in autistic neonates (Teitelbaum et al. 1998; Smith and Bryson 1998), as well as aberrations in emotional responses (Corona et al. 1998) and amygdala function, both of which are essential for social cognition and normal brain function (figure 5.8). We now turn to consider further the social brain.

Chapter 6
Social Reason and Action

The idea that the mind is detached from the body is modern and took on epistemological flair after the sixteenth century (Rorty 1979). In some respects, this idea was liberating; in other respects, it was a bad abstraction from a concrete fact (Whitehead 1927; James 1907). Although thought can be and often is expressed without acting, mind evolved in action or in imagining it or in anticipation of it. Although the two can obviously be separated, they are linked. In other words, the mind-body split is something pernicious on the one hand and an advance on the other. The ability to imagine action may have been aroused directly and been adaptive and selected for, or it may have been an outgrowth of other abilities that were selected for. Action and perception categories can be separated functionally, but they often run together in terms of neural processing (e.g., Lashley 1951; Jeannerod and Decety 1995; Marsden 1984a, b; Martin 1999; Decety et al. 1997).

I indicated in chapter 1 a variety of evidence in which imagining an event and actually seeing or hearing an event recruits many of the same neural systems. For example, in macaque monkeys, a neuronal population in the caudal region of the ventral premotor cortex was shown to be responsive both when the monkeys performed a particular hand-mouth movement and when they observed the experimentor doing the same (Gallese et al. 1996; Gallese and Goldman 1999). Motor systems are replete with motor images or information-processing systems (Annett 1996; Vogt 1996). In other words, the representation of possible movement is intertwined with thought (Jeannerod and Decety 1995).

Throughout the book, I have emphasized that the body is a vehicle for understanding (Clark 1996/1997; Heelan and

Schulkin 1998). The body is the medium by which we explore the world (Merleau-Ponty 1964), learn, and adapt to our surroundings (see also Varela et al. 1995; Johnson 1987; Lakoff and Johnson 1999; Heelan and Schulkin 1998). Perception is replete with cognition/representation, as is action (e.g., Merlau-Ponty 1942/1967, 1962/1994; Bogdan 1988, 1997; Clark 1996/1997; Shettleworth 1999; Harnard 1987/1990).

Representation of bodily events at the level of the forebrain is an integral aspect of the thought process (Damasio 1994). The root metaphor is the "embodied mind" (Varella et al. 1993; Lakoff and Johnson 1999; Clark 1996/1997; Merleau-Ponty 1942/1967, 1962/1994). The body is not something detached from cognition and action. In one tradition in biological psychology, information-processing systems are endemic to bodily experiences (e.g. Lashley 1938; Stellar 1954; Konorski 1967; Miller 1959; Panksepp 1998; Pfaff 1999; Schulkin 1999). Thus, for those of us who never made Descartes' error—namely, detaching reason and decision making from the body—can still appreciate Damasio's (1994, 1996) "somatic marker" hypothesis: the body in mind carves out the world to be understood. Damasio emphasizes the mind-brain assigning of positive and negative evaluation of events. This information-processing system in the brain operates quite effectively so that bodily knowledge informs rational thought; it does not necessarily detract from reason and can inform reason, as Damasio has pointed out.

In this chapter, consistent with earlier chapters, I argue for the characterization of the cognitive nature of the emotions—the cognitive basis of action and motor behavior. I further outline some of the underlying neural systems involved with both. Any full depiction of social sensibility and neural function will have to integrate conceptions of the emotions as information-processing systems, recognition of the experiences of others, and the organization of intentional action.

Cognitive Nature of the Emotions

One intellectual tradition envisioned emotion as on one side and cognition as on the other. Emotions were thought to be feelings, and feelings were a particular kind of sensation. Feelings were part of the body, and the body was separate from intellect (e.g., Descartes). This tradition of distinguishing emotion from cognition paralleled that of distinguishing sensation from cognition. For example, a number of theorists have asserted that sensory experiences pervade early experience and that emotions are linked to this relationship (e.g., Zajonc 1980; Panksepp 1998). No doubt that my eight-month-old child is replete with sensory information (putting everything in the mouth), but information processing pervades at each level of the neural axis, and therefore information processing and cognition are endemic to sensation and emotional systems (cf. DeSousa 1987/1995; Parrott and Schulkin 1993; LeDoux 1993). As Peirce (1992) noted, "we have no pure sensations, but only sensational elements of thought" (107).[10]

Another traditional account identified emotion as passions, an account that rendered the person passive. Something was done to or done in the person (see Sabini and Silver 1998 for a defense of this position). Reason, by contrast, was active, free, and so on. Keep reason pure was the dictum (e.g., Kant 1787). No less a rationalist than Freud held to this view and with a dominant sense that emotions were isolated and private (see also Brothers 1997). Emotions were tied to the irrational, the magical, the noncognitive (Spinoza 1668; Sartre 1948; Sabini and Silver 1998). This account, I believe, is broadly conceived and construes the emotions as a piece of pathology.

A cognitive/functionalist view, when broadened (Rey 1997), suggests that emotional judgments serve a role in behavioral adaptation (e.g., Parrott and Schulkin 1993; Lane et al. 1999). Emotions are information-processing systems; they prepare the animal to act and are part of the background in the organization of action (Frijda 1986; Gallistel 1990; though see Sabini and Silver 1998 for a very different view). Functionalists such as James (1907) or Dewey (1894, 1895) held this view. The emotional response in surprise games (peek-a-boo) with neonates and children is part of preparing for action amid the sense of uncertainty and releases tension. Dewey thought emotions such as grief prepared one to act in characteristic ways.

Emotions evolved to serve us. They are part of our adaptive problem-solving mechanisms (Darwin 1872; Davidson and Sutton 1995; Adolphs 1999b). They can be fast and accurate, and they can be self-corrective. They serve to interpret the environment and therefore are essential information-processing systems. They are not purely sensory. One thing that they do is to anchor us to or drive us away from others. They are essential for attachments critical for development (Bowlbly 1988; Panksepp 1998). They anchor us to the social world. They therefore tie us to each other's beliefs and desires and experiences.

Neuroscience of Emotions

Social emotions permeate our interactions with each other. We take them for granted. We look to discern the communicative emotions through their expression (Darwin 1872; Marler 1961, 1977; Duchenne 1862; Ekman and Davidson 1994). Disdain and disgust, joy and excitement are ways to express disapproval or approval of another person's actions (Sabini 1992). For example, I want you to know, at some times and at others not, of my disapproval or my approval. It is a judgment. You need to consider my mind and my actions. Displays are communicative, and they may or may not require that we note anything about the mental state. The information from such displays, to some extent, is used to predict behavior (Smith 1977). (See figure 6.1.)

Approach and avoidance behavioral responses underlie action in our social world. They underlie our connections and aversions to others. We are attracted to those things that we judge to be of value to us. We avoid those that we judge to the contrary. On the biological side, approach and avoidance have been characterized within learning theory (Miller 1971), in comparative biology (Schneirla 1966), and in the study of affect (Berridge 1990, 1999; Davidson 1992) and motivation (Stellar and Stellar 1985). Approach to and avoidance of objects that we readily use (Gibson 1966) involve the integration of motor and motivational systems (e.g., Nauta and Domesick 1978; Swanson and Mogenson 1981; Everitt, Cador, and Robbin 1989)—namely, the organization of action (Gallistel 1990; Kelley 1999).

Figure 6.1
Facial communicative expressions (Darwin 1872/1965).

There appear to be two broad systems in the brain that underlie approach to and withdrawal from objects. The representation of goal objects by hedonic judgments is linked to positive affect and approach and to negative affect and withdrawal (e.g., Schneirla 1966; Konorski 1967; Fox and Davidson 1988; Berridge 1999; Stellar and Stellar 1985; Bradley and Lang 1999). In humans, the results of one study using PET and showing positive versus negative pictures of events demonstrated preferential activation of the left prefrontal cortex and striatum with positive pictures (Davidson and Rickman 1999). When negative pictures were shown, which would have elicited withdrawal, there was preferential activation within the right prefrontal cortex (Brodmann area 9) and amygdala.

What are the neural structures that make social hermeneutics possible? First, I would suggest that they are structures common to other functions that we perform. Second, they are structures linked to emotion or motivation and action. There is evidence within the theory of mind literature of the important link between understanding mind and understanding emotion (e.g., Brothers 1990, 1997). Regions of the brain that are fundamental in emotion should also be involved in the attribution of beliefs and desires to others. Regions of the brain that include the amygdala, basal ganglia, and the prefrontal cortex are involved in the organization of intentional behavior to approach or avoid objects (e.g., Everitt, Cador, and Robbin 1989).

The Amygdala, Emotions, and Social Cognition

The amygdala (figure 6.2) is old cortex housed in humans within the anterior region of the temporal lobe. It is an almond-shaped nucleus (Herrick 1905; Johnston 1923; Papez 1929). Regions of the amygdala receive direct connections from brainstem sites such as the solitary nucleus, parabrachial region, and the vagal complex; the neural connectivity is bilateral (e.g., Schwaber et al. 1982; Norgren 1976; Gray and Magnuson 1987). These regions of the brainstem are fundamental to visceral regulation. In addition, connectivity between the hypothalamus and

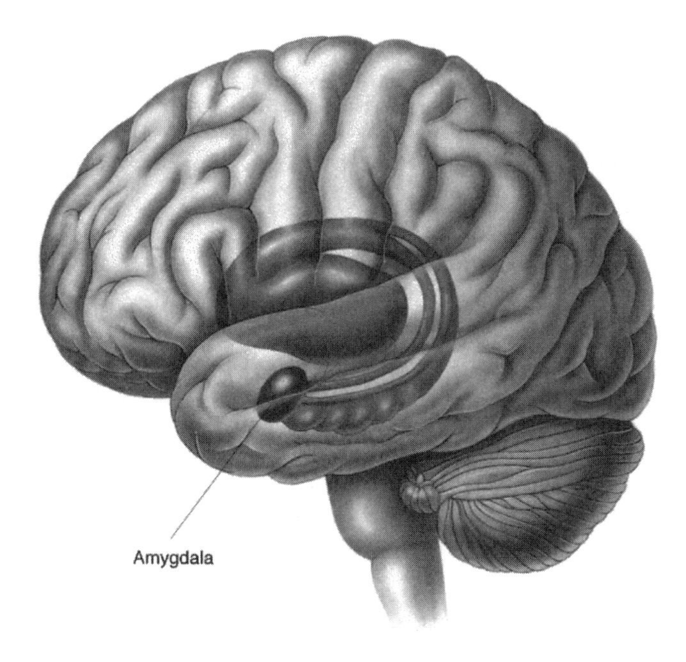

Figure 6.2
Picture of the brain with the amygdala highlighted (after McEwen and Schmeck 1994).

the amygdala are prominent (LeDoux 1987, 1996; Kretek and Price 1978; Swanson and Petrovich 1998). There are massive projections to the amygdala from neocortical and thalamic sites via the lateral nucleus (LeDoux 1996), and there is direct connectivity to the basal ganglia (Nauta and Domesick 1978; Alheid, deOlmost, and Batramino 1996; Swanson and Petrovich 1998).

The amygdala was once viewed as the smell-and-taste brain (Herrick 1905; Papez 1929; Norgren 1995). Indeed, regions of the amygdala are known to be involved in a number of basic regulatory behaviors in which smell and taste play important roles in the expression of the behavior (Schulkin 1999; Norgren 1995).

The amygdala is importantly involved in the regulation of the emotions, particularly fear (LeDoux 1987, 1996; Fonberg 1974;

Aggleton 1992), but it has a much larger role to play in the regulation of behavior, specifically in the attention to and evaluation of positive and negative information (Gallagher and Holland 1994), and perhaps in the approach to or avoidance of objects. It is linked to the assessment of facial responses (e.g., Rolls and Treves 1998; Adolphs et al. 1999). The amygdala's role in emotional judgment involves anticipation of future events (Schulkin, McEwen, and Gold 1994).

The amygdala has been linked to social communication in a number of species (e.g., Bamshad et al. 1997; Fonberg and Kostarczyk 1980; Brothers 1997). It has long been known that damage to the amygdala can affect an animal's interpretation of social events (Schreiner and Kling 1953). The classic Kluver-Bucy (1939) syndrome reflected the finding that bilateral amygdala damage resulted in cats inappropriately performing sexual acts. They show decreased contact in addition to misinterpretation of conspecifics. In general, lesions of the amygdala seem to decrease interpretative competence and to blunt affect (e.g., Fonberg 1974; Fonberg and Kostarcyzk 1980; Weiskrantz 1956).

I indicated in the previous chapter that the amygdala is now linked to autism (Baron-Cohen et al. 1999). With regard to social behavior, it is typically the case that in the nonlaboratory setting and semiwild settings, amygdala damage results in social isolation and loss of social contact across a wide variety of species. These effects are elegantly summarized in a chapter by Kling and Brothers (1992; see also Brothers 1994; Emery and Amaral 1999).

Neurons within the amygdala and in the temporal cortex are responsive to a broad range of social stimuli, including eye position (e.g., Brothers et al. 1990; Perrett and Emery 1994; Rolls and Treves 1998). Neurons within the amygdala are responsive to movement sequences and to novel objects in the environment (Brothers and Ring 1993; Rolls and Treves 1998).

One study using fMRI to measure brain activation (figure 6.3) demonstrated greater activation of the amygdala when a subject sees faces that are frightening compared to those that are not

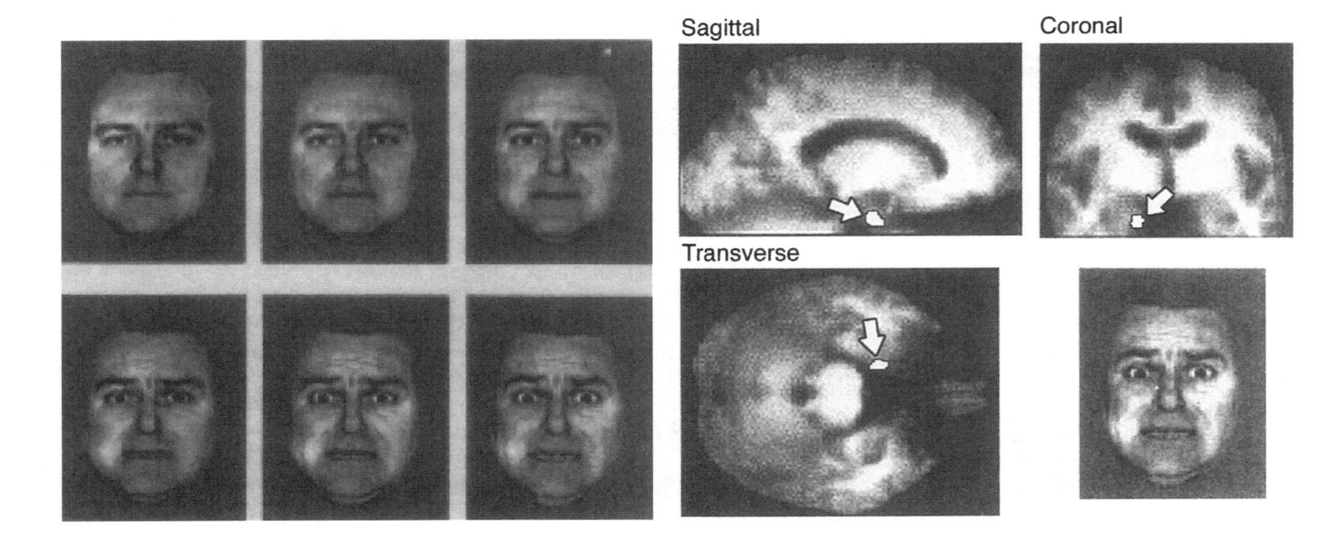

Figure 6.3
Facial expressions from neutral to fearful. Activation of the left amygdala (arrows) in subjects shown photographs of fearful faces (Morris et al. 1996).

(e.g., Morris et al. 1996; Breitter et al. 1996). Of course, the amygdala is not alone in the processing of facial information; different regions of the brain are involved in different determinations of facial information processing (e.g., Haxby et al. 1999).

Evidence suggests that patients with complete bilateral lesions of the amygdala approach unfamiliar objects more readily than control subjects (Fonberg 1974; Adolphs, Tranel, and Damasio 1998). Patients with complete bilateral damage of the amygdala are unusual with regard to their responsiveness to faces of unfamiliar people in two contexts: first, they are more willing to approach them, and second, they are more likely than controls to rate them as more trustworthy. (See figure 6.4.)

Recent studies have shown quite convincingly the role of the amygdala in normal fear (LeDoux 1996) and perhaps in aberrant fear (Rosen and Schulkin 1998). We have shown, for example, that in rats in which the amygdala has been kindled (electrically stimulated) weeks prior to a fear-conditioning context, the result is a greater fear-related response (e.g., startle). The prior activation of the amygdala resulted in a propensity to perceive an event as fearful (Rosen and Schulkin 1998). But the circuitry within the amygdala is recruited for other tasks, including social ones (Brothers et al. 1990).

Basal Ganglia and the Neuropsychology of Action

The amygdala's connectivity to the striatum represents the connection between motivation and the expression of action patterns (Nauta and Domesick 1978; Swanson and Mogenson 1981; Everitt, Cador, and Robbin 1989; Kelley 1999; McGinty 1999). Regions of the brain that include the amygdala, basal ganglia,

Figure 6.4
Mean approachability and trustworthiness rating for the fifty most negative faces and the fifty most positive faces. Subjects with bilateral amygdala damage or control subjects judged unfamiliar people on approachability and trustworthiness. Unlike the control subjects, those with bilateral amygdala damage rated negative faces as trustworthy and approachable (Adolphs, Travel, and Damasio 1998).

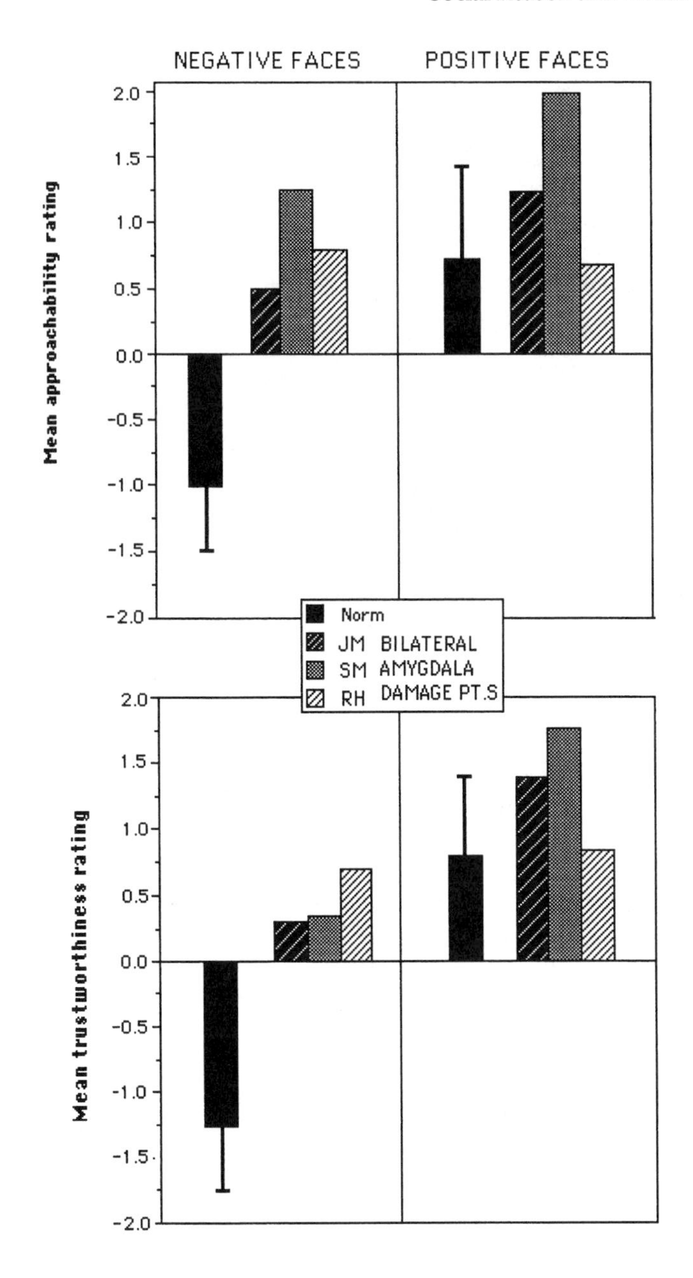
NEGATIVE FACES
POSITIVE FACES
Mean approachability rating
Mean trustworthiness rating
2.0
1.5
1.0
0.5
0.0
-0.5
-1.0
-1.5
-2.0
Norm
JM
SM
RH
BILATERAL
AMYGDALA
DAMAGE PT.S

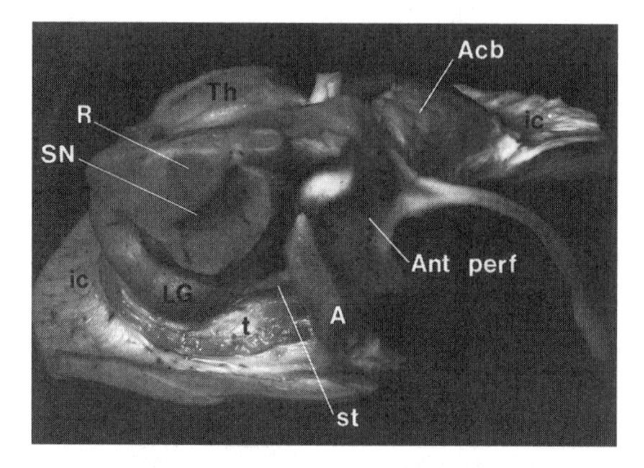

Figure 6.5
Ventral view of the basal ganglia of the human brain. Note the expansion of the tail of the caudate as it courses rostrally into the temporal lobe. The rostral half of the amygdala (A) has been cut away to better expose the anterior perorated space. The substantia nigra (SN) has been exposed by a transverse section of the mesencephalon at the level of the superior colliculus. The olfactory bulb and lateral tract remain attached to the base of the brain, but are displaced laterally to expose the accumbens (Acb) and the anterior radiation of the internal capsule. Abbreviations: Ant perf, anterior perforated space; R, red nucleus; Th, thalamus; st, stria terminalis (Alheid and Heimer 1996).

and the frontal cortex are fundamental in the organization of action (Graybiel 1998; Rolls and Treves 1998).

The basal ganglia is a large region of the brain with a number of nuclei (Johnston 1923; Alheid and Heimer 1996; Brodal 1969/1981). Within the basal ganglia (figure 6.5) are the striatum, globus pallidus, substantia nigra, and several subthalamic nuclei (Rolls and Treves 1998; Alheid and Heimer 1996; Swanson and Petrovich 1998). In some anatomical descriptions, regions of the amygdala are also included within this region.

A specific role of the basal ganglia is in the organization of movement (Marsden 1984a, b; Graybiel 1997; Kelley 1999; Berridge 1999). Dopamine, a neurotransmitter that underlies a number of behavioral functions including that of voluntary movement, is heavily innervated and synthesized within the basal ganglia (Marsden 1984a, b; Holt, Graybiel, and Saper

1997). Interestingly, Parkinson's patients, who have decreased capacity to synthesize dopamine, also reveal not only decreased ability to generate real movements, but also decreased speed in imagining movements (Dominey et al. 1995).

The basal ganglia orchestrates the sensorimotor integration involved in movement sequences (Graybiel 1997; Kelley 1999). This region of the brain that underlies species-specific behavioral sequences may also underlie procedural-computational or rule-governed behavioral sequences (e.g., procedural or habit memory; Mishkin, Malamut, and Bachevalier 1984]). It has been argued, I think erroneously, that the striatum, within the basal ganglia, that underlie action or the memory of action is noncognitive (e.g., procedural knowledge; Mishkin, Malamut, and Bachevalier 1984]). A distinction on conceptual and then anatomical grounds between declarative and procedural memory is a real one, and habits, or procedures, are for the most part outside of consciousness (e.g., Squire 1987, 1998). But cognition pervades both declarative and procedural memory. "Procedural knowledge"—knowing how to fix something (Ryle 1949)—is replete with cognition, no matter how habit forming it appears, and the cognitive mechanisms, like others, may be unconscious to us—with no privileged access. Cognition is not on one side and action on the other.

In fact, no action is simply sensorimotor (Dewey 1896; Lashley 1951; Bogdan 1997; Lakoff and Johnson 1999). Both emotions and movements are replete with cognition (see also Bogdan 1997; Graybiel 1998), and certainly anything we would call action is also replete with perception.

Action programs are a feature of the brain (Jeannerod 1994; Graybiel 1997). Lashley (1951, 1958) thought that movement revealed a form of syntax. Berridge and his colleagues have discovered levels of the brain and the form of expression that can be understood in terms of syntactic-movement structure (Berridge 1990; Berridge, Fentress, and Parr 1987). Movement sequences in rat facial and hand movements, for example, can be understood as chains of rules generated by regions of the brain (striatum) that underlie action and the approach and avoidance of objects (Berridge 1990). It is still nothing like the syntax of language and should not be confused with language. But action repertoires are inherently cognitive and are mediated by the basal ganglia (striatum; Graybiel et al. 1994; Knowlton, Mangels, and Squire 1996).

A number of investigators (Damasio 1996; Ullman et al. 1997; Graybiel 1995; Marsen 1984a,b) have suggested that the basal ganglia in addition to the frontal cortex underlie motor programs that range from movements to grammatical rule processing. These regions of the brain subserve both specific and general functions. The basal ganglia is linked to wide variety of cognitive events (probabilist reasoning, artificial grammar; Squire and Zola 1996), but all are within the confines of rule following.

Consider the paradigmatic cognitive function of language. Some evidence suggests that patients with basal ganglia damage have difficulty with specific forms of grammatical processing (Ullman et al. 1997). For example, Parkinson's patients have difficulty with specific grammatical rules: the greater the motor deficit, the greater the decrement in rule-based regular versus irregular memory-based verb production. These patients, again, in general have lower levels of dopamine and a wide array of both motor and cognitive impairments. Moreover, such grammatical rules have been linked to procedural memory systems that include the basal ganglia (Ullman et al. 1997). (See figure 6.6.)

A PET study revealed that normal subjects show greater activation for regular versus irregular verb production in the left prefrontal cortex (Jaeger et al. 1996). A further study using fMRI demonstrated greater activation of the caudate nucleus within the basal ganglia in the production of regular than irregular past-tense forms (Ullman et al. 1997). These results suggest that the basal ganglia and frontal cortex are essential in the production and use of grammatical rules. (see also Joanisse and Seidenberg, 1999).

Finally, one should also consider the observation that aberrations of both thought and movement disorders (e.g., obsessive compulsive disorder) have been linked to both an overactivation of the striatum and frontal cortex, as has Tourrette's syndrome (Wright et al. 1999; Swerdlow and Young 1999). Much of this discussion gives credence to James's claim that the habits of thought and the habits of movement are not very distinct from one another (e.g., Graybiel 1997; Clark 1996/1997).

Knowledge of Others, Visceral Anticipatory Responses, and the Frontal and Temporal Cortex

The hypothesis throughout this book is that social knowledge, in part, is the attribution of intentions, or beliefs and desires, to others. These events have their origins in predicting the behavior of others in action in the social world. The attribution of beliefs and desires to others is also one way in which we come to learn about other people's experiences. Its roots are within biological adaptation.

In the last chapter, I discussed the findings regarding autistic children, which suggested decreased frontal cortex activation to perceived intentional social events when compared to controls. This region is known to be important in a number of cognitive functions—in particular, social competence (e.g., Fox et al. 1995; Gazzaniga 1995; Damasio and Maurer 1994), so we should not be surprised that it is linked to the recognition of the beliefs and desires of others.

The frontal cortex is the largest structure of the neocortex and of the brain in general. It comprises nearly a third of the brain. Regions of the frontal cortex underlie memory and a wide array of mental functions (see, e.g., classic papers in Von Vonin 1960; Herrick 1926; Goldman-Rakic 1996; Smith and Jonides 1999), including the emotions (Davidson 1992; Adolphs 1999b), social ability (Damasio, Tranel, and Damasio 1990; Luria 1973), and bodily sensibility (e.g., Mountcastle 1978, Damasio 1996) and motor control (Graybiel 1995).

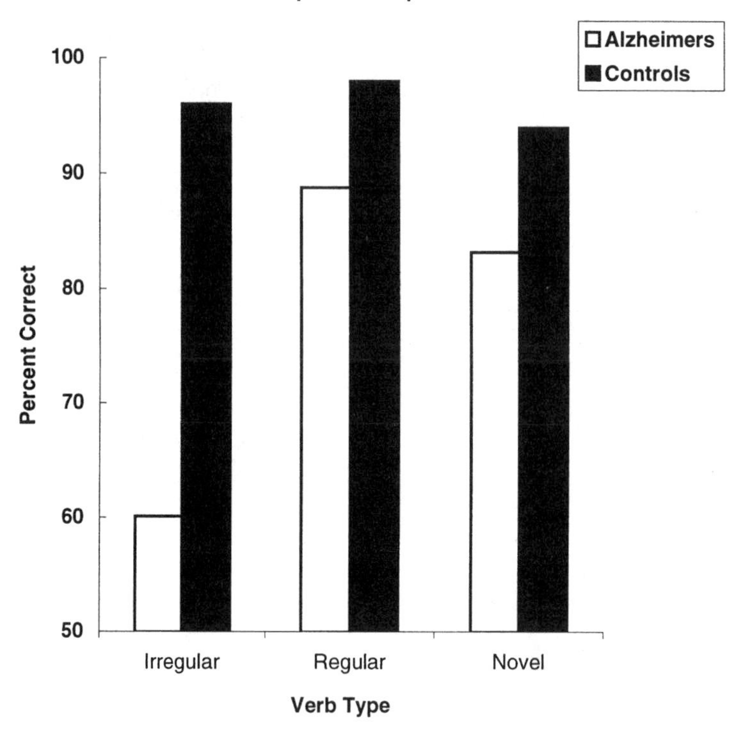

Figure 6.6
Percent correct responses among patients with Alzheimer's disease (left) and Parkinson's disease (right) compared to controls when asked to produce past tenses of irregular, regular, and novel verbs. Parkinson's disease patients showed suppressed use of grammatical rules, whereas Alzheimer's disease patients showed normal rule use, underscoring a role for the basal ganglia in grammatical processing (Ullman et al. 1997).

It is apparent that this region is important for understanding something about other people's experiences. One experiment using EEG to measure cortical brain activity demonstrated greater left frontal activation in normal subjects when they were forced to represent false beliefs about an object rather than represent the object itself (Sabbagh et al. 2000).

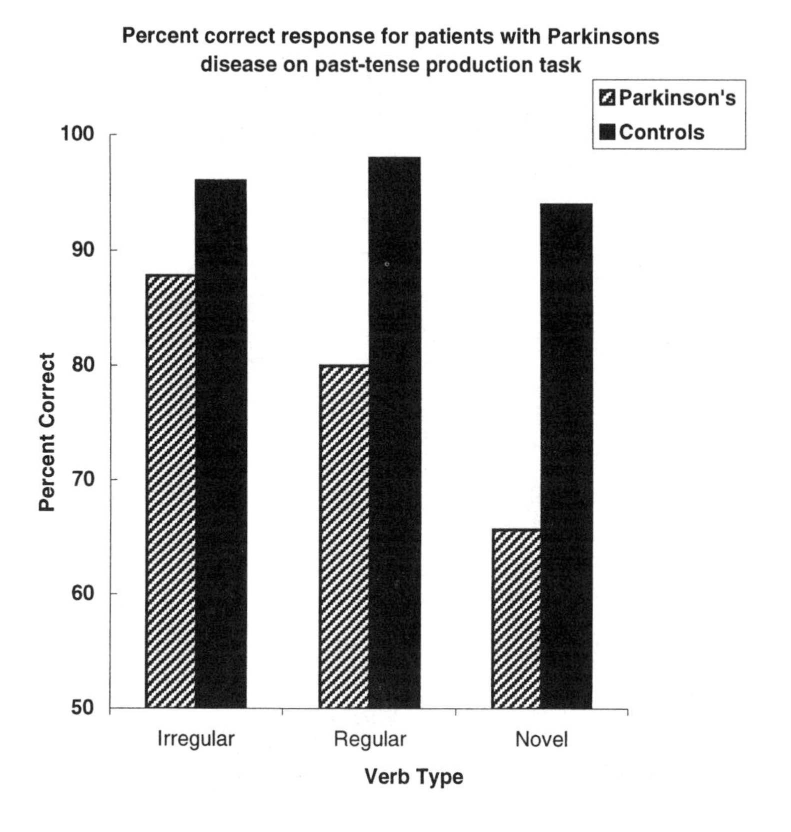

Another study using PET demonstrated greater regional activation of the frontal cortex (Brodmann area 9) when normal subjects were forced to draw inferences about the beliefs and desires of others than when other mental functions were used (Goel et al. 1995). Other regions activated were various regions of the temporal lobe (Broadmann areas 21, 39/19, 38). (See figure 6.7.)

Bilateral damage to the orbito-frontal region comprises social reasoning (Stone, Baron-Cohen, and Knight 1998). Subjects with either bilateral damage to the orbito-frontal region or unilateral damage the left dorsal lateral region were tested. Bilateral orbito-frontal subjects performed similar to Asperger syndrome patients in a faux pas context (saying something that one should

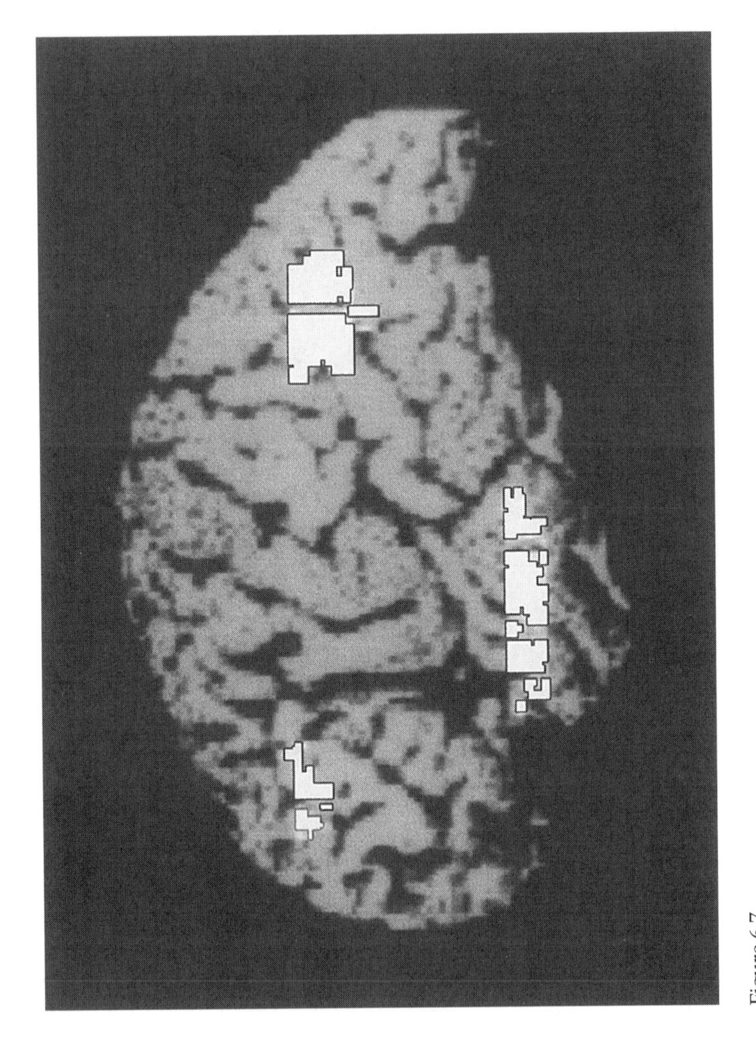

Figure 6.7
Activation of regions of the brain (light areas) when forced to think about the minds of others (Goel et al. 1995).

not have and noticing how it affects others). Both had difficulty on a version of this test. Clearly the frontal cortex is involved in social knowledge or social sensibility, but it is involved in a vast variety of cognitive functions as well (Milner and Petrides 1984; Adolphs 1999).

More generally, impairment in planning ahead is always one of the most striking features in individuals with prefrontal damage (e.g., Luria 1973; Goel et al. 1997), and perhaps the recognition of planning in others is also something disrupted in patients with such frontal cortical damage. Thus, as shown in one example regions of the human prefrontal cortex underlie decision making linked to rewards and punishments (visceral) when future events are considered (Bechara et al. 1996, 1997), which is reflected by autonomic activation in normal subjects. Patients with prefrontal cortical damage do not show autonomic conductance responses that reflect anticipatory responses (Bechara et al. 1996, 1997). (See figure 6.8.) After all, even imagining grasping an object is known to activate the visceral or autonomic system (Decety, Durozard, and Baverel 1993). On the other hand, Asperger patients who showed decreased performance in theory of mind tasks revealed no difference in the activation of the primary motor cortex when viewing hand movements in so-called "mirror neurons" in this region of the brain (Auikainen et al. 1999).

Visceral responses at the highest level of the cortex are important for human decision making (Damasio 1994, 1996). In other words, implicit in decision making is a visceral representation at the level of the brain that helps structure action and movement to approach or avoid objects and that are informed by these events (Rolls and Treves 1998; Rizzolatti and Arbib 1998; Jeannerod 1999).

Finally, patients with prefrontal damage, particularly within the ventral and medial regions, display decrements in social reasoning but near normal performance in a number of other cognitive functions (Damasio 1994; Tranel et al. 1994). Regions of the frontal cortex serve as an important source of our visceral decision making.

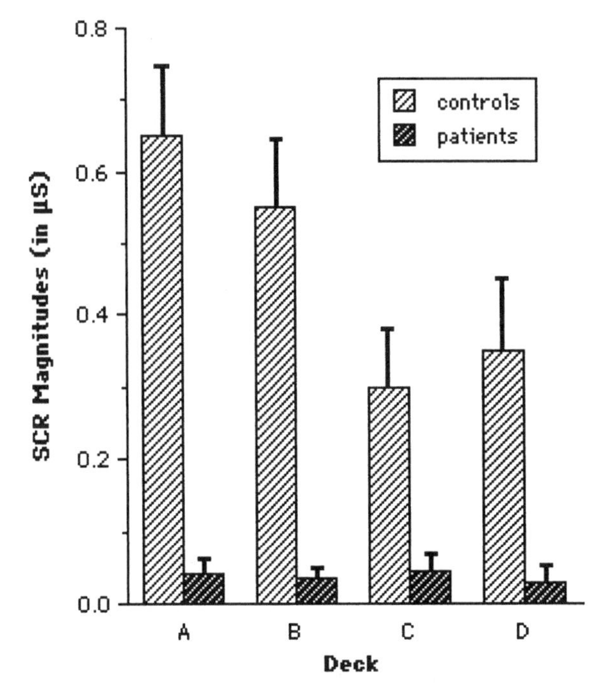

Figure 6.8
The magnitude of anticipatory skin conductance responses (SCRs) generated by controls and by patients with damage of the prefrontal cortex (Bechara et al. 1996).

Summary

Information-processing systems pervade the emotions and the organization of action. Neural circuits underlie action, perception, social emotions, and motivation—all of which are replete with cognition. Regions of the frontal and temporal cortex, basal ganglia, and amygdala underlie the organization of action and perhaps our ability to parse our social space (Adolphs 1999). Action programs permeate the parsing of our social space. On the analogy with the visual system, perhaps viewing intentions in others, imagining them in ourselves or others, may recruit the same core brain regions in the organization of action. In other words, the recognition of others and many other cognitive events underlie action; action programs are replete with cognition. Therefore, one would expect that intentional action and the perception of it in others would be embodied in a circuit that includes the frontal and temporal cortex, basal ganglia, and amygdala.

Conclusion

In a previous book, I suggested that social knowledge or social intelligence, along with language use, was perhaps the most distinguishing feature about us. I am certainly not alone in holding this view, but I also suggested that intentionality was a precondition of social knowledge and that social knowledge was key in understanding the experiences of others (see also Neville 1974; Smith 1970, 1978; Schulkin 1992). The use of this knowledge probably evolved in the context of predicting the behavior of conspecifics and was extended in use to a variety of domains. The use of knowing other people's beliefs and desires was important for keeping track of them—bonding, deceiving, manipulating, and the whole range of other possible human interactions.

In this book, I have reinforced and expanded on a neural framework (e.g., Perrett and Emery 1994; Rolls and Treves 1998; Brothers 1990; Baron-Cohen et al. 1999) that may underlie our understanding of the beliefs and desires of others—of their experiences. Regions of the amygdala, frontal and temporal cortex, and basal ganglia may underlie this ability to understand others people's intentional behavior and may be active when a person imagines intentional action.

Importantly, neurodevelopmental pathology, genetically based, may underlie the failure to develop this ability (e.g., autism). In this regard, transgenic models that delete or enhance neuropeptides in the brain have been shown to alter behavior. For example, transgenic models, in which oxytocin gene expression in the brain is altered, affect affilitative behaviors (Insel 1997; Young et al. 1997). Genetically engineered models in which oxytocin in mice has been deleted demonstrate impairments in social behavior (Nishimori et al. 1996). More generally, central infusions of oxytocin increase the duration of time spent in friendly bodily contact (see also Carter et al. 1997; 1999). Moreover, increases in affiliative behaviors can occur by altering central vasopresin receptors, a structurally closely related neuropeptide (Young et al. 1999) Insel and his colleagues (1999) have suggested that genes that code for oxytocin or a subset of the vasopressin receptor (Vla) may play an important role for some aspects of autism.

Moreover, perhaps differences in the behavior of the bonobo chimpanzee (i.e., the use of social tranquility; see figure C.1. and DeWaal and Lanting 1997) from that of the common chimpanzee may reflect differences in neuropeptide oxytocin gene expression in the brain (e.g., amygdala, extended amygdala) (Insel 1993; Carter et al. 1999).[11] In other words, aggression may be reduced by forming attachments that reflect genetic differences in oxytocin production in functional circuits of the brain. It is at least a testable idea.

In fact, a wide diversity of neuropeptide hormones are found in these (amygdala, bed nucleus of the stria terminalis) regions of the brain (Herbert 1993; Schulkin 1999), and they have been suggested as playing a role in autism (Panksepp 1998; Insel et al. 1999). For example, lower levels of oxytocin have been noted in autistic children than in normal children (Modahl et al. 1998). Plasma levels of oxytocin (figure C.2) reveal modest differences between autistic and normal children. Of course, this is peripheral oxytocin and what we really want to know are central levels.

Intentionality and Social Parsing, Opaque and Real

The design of minds like ours is essentially interpretative (Hiley, Bohman, and Shusterman 1991), replete with a rich sense of experience (Flanagan 1996; Haugeland 1998). How to capture that experience is still the great puzzle of inquiry. The cognitive and neural revolution needs to be anchored to a conception of experience that is active—a view of perception that is anchored in

Figure C.1
Bonobos in one setting (DeWaal and Lanting 1997).

human action, adaptation, and the search for coherence. The fact that seeing is always perspectival—whether for the child, other animals, or investigators—does not denigrate the matter into idealism or entrap it in the depiction of cut-off mind.

Intentional relations permeate our discourse with one another, perhaps influenced by both local habitat and biological

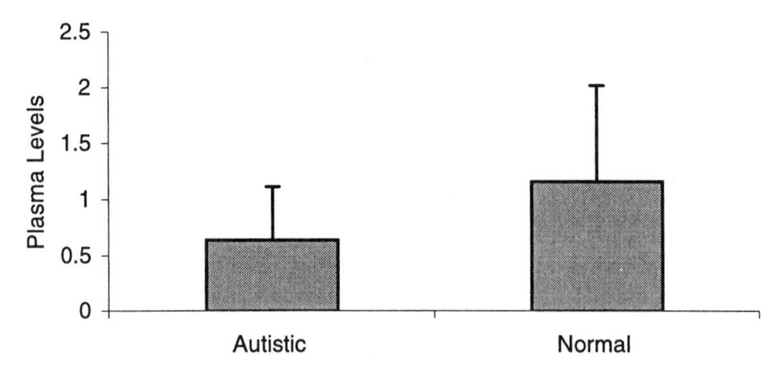

Figure C.2
Oxytocin levels in autistic and normal children (Modahl et al. 1990).

predilection (Hughes and Cutting 1999). Moreover, intentional attributions come easily, perhaps in some instances too easily. The ability to distinguish between truly intentional behavior and unintentional behavior is a performance issue and will vary among individuals and with experience. Thus, it is endlessly problematic whether behavior may or may not be intentional (e.g., deception studies, pretense studies, etc.). In addition, there is no hard rule that axiomatically dictates that something reveals something about intentions to the actors themselves—there may be nothing in mind (cf. Joseph 1998; Lillard 1998). Yet it is useful for us to predict behavior from the intentional stance (Dennett 1969, 1987).

Of course, from the realist philosophical point of view, we want to know when something is really intentional and when it is not. I submit that considerations, perhaps like a number of other cognitive functions, that we attribute to others (e.g., motivation) will always be somewhat opaque. Perhaps more important are the recognition of the amazing abilities that are revealed about the mind and the continued search for the experiments that probe deeper into these abilities and the brain mechanisms that render them possible. The idea that folk psychological discourse is itself a theory of human and animal behavior is justified, but it must be viewed with caution (e.g., Hauser and Carey 1998; but cf. Churchland and Churchland 1998).

We often learn by looking out on the world and interacting with others, in addition to reflecting on our own beliefs and desires (Wittgenstein 1953; Peirce 1898; Gadamer 1975). Social knowledge is bound to communities (e.g., Peirce 1889; 1992; Wittgenstein 1953; Goffman 1959) and not simply isolated within individuals. Knowledge is not a spectator sport (Dewey 1925), but an active form of engagement. The problem with the metaphor of the "introspective mind" (Humphrey 1974) is that historically it was a barrier to embodying the mind in the community and to legitimating knowledge in the lone individual, whether in empiricist or rationalistic terms. Social hermeneutics is oriented to the world first and foremost (e.g., Mead 1938; Sabini and Silver 1982, 1998; Bogdan 1997).

For our species, knowledge is intricately linked to social practice; knowledge of others and what they know is passed on to us in communities, which Peirce called "the community of inquirers." The intellectual stage is set for at once acknowledging social constructivism and realism (Sabini and Schulkin 1994). Knowledge is not private in any epistemological sense (Wittgenstein 1953; Gadamer 1975), and those experiences of knowledge and practice are shared (Goffman 1959; Geertz 1973).

In this regard, an interesting fact about us as a species is the extent to which all our cultural communities talk about events in terms of beliefs and desires (cf. Lillard 1998; Wellman 1998). In fact, I suspect that a systematic exploration of the way in which subcultures of dominant cultures go about explaining the intentional actions of their conspecifics will show great variation. There is great variation in language, but that does not mean that the mechanisms for language production are not the same (Chomsky 1972; Pinker 1994, 1997b).

The cognitive scientist needs constraints, and those constraints can come from two directions, from the biology of the organism and the social context in which the organism must function. It is enough for the investigator of artificial intelligence (e.g., Marr 1982) to invent some process by which some specific

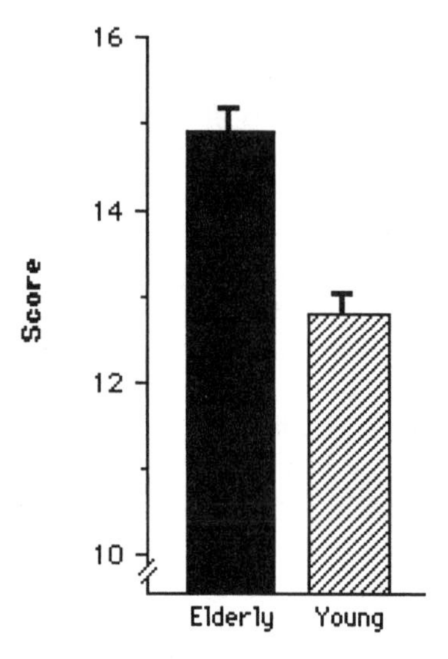

Figure C.3
Elderly and young people's response score to theory of mind stories (Happé et al. 1998).

outcome is accomplished, but the cognitive scientist wants to know not only how might *x* be accomplished, but how we humans (or perhaps other organisms) do it. For that, biology is often needed. Cognitive scientists have a habit of ignoring the social context and of imagining that the phenomena that form culture are only the upshot of internal, cognitive processes.

Final Thoughts

It is interesting that in one study, older people performed better in tests that required making inferences about beliefs and desires of others than did young adults. When they were asked to judge a story that did not require inferring beliefs and desires to others, the subjects in both groups performed the same (Happé, Winner, and Bronell 1998). Some of the data from that study are depicted in figure C.3.

The authors of this study thought that older folks had "superior social insight"—at least these subjects did. Perhaps there is something to the ideas that age (for a subset of us) can bring on wisdom and that wisdom is, in part, understanding the experiences of others.

Amid a sense of social action replete with human intention is an appreciation of other people's experiences. The knowledge of these experiences, encompassing a knowledge of the various beliefs and desires of others, is set in the context of our social world. The social milieu—those everyday structures that organize our lives—pervades our every intentional move. Within the diverse expressions of our everyday lives is a normative goal: by understanding the experiences of others we may act more favorably toward one another.

"But all things are as difficult as they are rare" (Spinoza 1668, p, 271).

Notes

1. Peirce is noted as developing the first experimental psychological laboratory in the United States (see T. D. Cadwaller's 1974 article "C. S. Peirce: The First American Psychologist"). The pragmatism that pervades this book is one in which social action is tied to inquiry, a sense of nature, common-sense realism, and the learning about other people's experiences (Smith 1978; Neville 1974, 1989; Sleeper 1986). This is in contrast to more popular revisionist gesticulations (Rorty 1982, 1991), which I fear have once again given pragmatism a bad name. Rorty's version of science is just more talk amid other talk. Moreover, he does not retain the notion of inquiry dear to pragmatist sensibility, which is in contrast to the scientism that emerged in the writings of Quine (e.g., 1961, 1966, 1969). Elsewhere I (1992) have argued against the Richard Rorty view of pragmatism that eschews the early pragmatist sense of biological science and realism (see also Mounce 1997). On the broader sense of community, early pragmatists always emphasized the inherent meaning of our social interactions. This emphasis is poignant in pragmatists such as Dewey (1910, 1916, 1925/1989; see Smith 1970, 1978) and Mead (1932, 1934; see Joas 1993, 1996).
2. The issue is that in one case there really is a fact of the matter and in the other there need not be. The latter result leads some to want to leave intentional discourse behind in the future as our epistemology advances (Churchland and Churchland 1998). Realists, like myself, want to hold onto the intentional level of analysis and locate it in the brain, along with a language of intentional description (see Rey 1997, 1999).
3. The idea of motion or action figures importantly in the chapters ahead with the idea of intentional action (see, for example, Premack 1990; Perrett and Emery 1994; Hauser and Carey 1998).
4. Consider domestic male chickens. They usually produce a distinct call when they discover food, and the rate of the call increases with the palatability of the food. Whether or not the chickens produce calls depends on their audience. A series of experiments evaluated how male chickens responded to food when the cage next to them contained one of the following: a familiar female chicken, an unfamiliar female, nothing, or an adult male (Marler, Dufty, and Pickert 1986a; Marler 1990a). When a female, familiar or not, was next

door, the male chickens almost always called when they discovered food; they sometimes called (more often than not) when there was nothing in the cage next to them; and they almost never called when there was a male in the next cage. Viewing this phenomenon from an evolutionary perspective, it is to the chicken's benefit to attract females with the call of food, but to withhold that information from another male so that he can have more food to himself. This discrimination demonstrates functional withholding of information (Trivers 1985; Dawkins and Krebs 1978) so paradigmatic of game theory's conception of human action (Axelrood 1984; Elster 1979/1988). See also Pepperberg (1990) for a discussion of the conceptual abilities of nonprimate species.

5. A great debate is being waged on the use of folk psychology and whether it is legitimate (e.g., Stitch 1983; Churchland and Churchland 1998). As a pragmatist, I have no doubt of its use, but also of its abuse. So one looks for some restraint on the reflexive use of mentalism and some restraint on excessive reductionism, or the idea that our species and limited experimental paradigms necessarily eliminate the legitimacy of this concept. I do not think we know yet. Of course, I am reminded of a colleague and friend (John Sabini) who would probably put a moritorium on debates like this.
6. Chimps do not appear to have anything like the socialization of attention we do (Tomasello and Call 1997, though see Whiten et al. 1999). But who would expect them to? And our learning capacities are outrageously beyond our closest ancestors, though variation in their behavioral patterns can be quite impressive (see Whiten et al. 1999).
7. The concept of preadaptation (see Mayr 1969; Rozin 1998)—namely, the use of a function that it did not evolve for or the extension of a cognitive adaptation to a novel domain—may underlie our ability to attribute beliefs and desires to others.
8. This issue about whether the child is a hypothesis tester—or self-corrective, like a miniscientist—and the debate around it is somewhat analogous to the debate around Jennings 1905/1962 and Loeb 1918/1973 and whether lower animals are hypothesis testers or trophic reflex machines. Children have been characterized as little scientists testing hypotheses. I think the important thing is that it is recognized that children have in their mind-brain hypotheses or theories that guide their behavioral responses and the judgments of other people's behavior.
9. What is contrasted is sensation versus cognitive disorders with regard to autism. But if we, as I do, reject "the myth of the given" by suggesting that all seeing is seeing relative to a background framework (Sellars 1963, 1968), the distinction between the sensory and the cognitive (chapter 2 and 6; cf. Parrott and Schulkin 1993a, b; LeDoux 1993, 1996; Zajonc 1980) is not the legitimate way to bifurcate the differences here.

10. Although the issue is complicated, I see no reason to suggest, as some have, that qualia are "first order properties of represented intentional objects" (Lyons 1995, p. 160), for surely something can be a quality without it being intentional. Qualia are, however computed, represented in a background framework.
11. This idea emerged in the context of a telephone conversation with Jerry Kagan (1998).

References

Abercrombie, H. C., Schaefer, S. M., Larson, C. L., Oakes, T. R., Lindgren, K. A., Holden, J. E., Perlman, S. E., Turski, P. A. Krahn, D. D., Bencea, R. M., and Davidson, R. J. (1998) Metabolic rate in the right amygdala predicts negative affect in depressed patients. *NeuroReport* 9: 3301–3307.

Adolphs, R. (1999a) Neural systems for recognizing emotions in humans. In *The Design of Animal Communication*, edited by M. D. Hauser and M. Konishi. Cambridge, Mass.: MIT Press.

Adolphs, R. (1999b) Social Cognition and the human brain. *Trends in Cognitive Science* 3: 469–479.

Adolphs, R., Damasio, H., Tranel, D., and Damasio, A. R. (1996) Cortical systems for the recognition of emotion in facial expressions. *Journal of Neuroscience* 16: 7678–7687.

Adolphs, R., Tranel, D., and Damasio, A. R. (1998) The human amygdala in social judgment. *Nature* 393: 470–474.

Adolphs, R., Tranel, D., Damasio, H., and Damasio, A. (1994) Impaired recognition of motion in facial expressions following bilateral damage to human amygdala. *Nature* 372: 669–672.

Adolphs, A., Tranel, D., Hamann, S., Young, A. W., Calder, A. J., Phelps, E. A., Anderson, A., Lee, G. P., and Damasio, A. R. (1999) Recognition of facial emotion in nine individuals with bilateral amygdala damage. *Neuropsychologia* 37: 1111–1117.

Aggleton, J. P., ed. (1992) *The Amygdala: Neurobiological Aspects of Emotion, Memory, and Mental Dysfunction*. New York: Wiley-Liss.

Ainsworth, M. D., Blehar, M. C., Waters, E., and Wall, S. (1978) *Patterns of Attachment*. Hillsdale, N.J.: Lawrence Erlbaum.

Aldridge J. W., and Berridge K. C. (1998) Coding of serial order by neostriatal neurons: A natural action approach to movement sequence. *Journal of Neuroscience* 18: 2777–2787.

Alheid, G. F., deOlmost, J., and Batramino, C. A. (1996) Amygdala and extended amygdala. In *The Rat Nervous System*, 2d. ed., edited by G. Paxinos. San Diego: Academic.

Alheid, G. F., and Heimer, L. (1996) *Basal Ganglia: The Human Nervous System*. Edited by George Paxinos. San Diego: Academic.

Allen, C., and Bekoff, M. (1997) *Species of Mind: The Philosophy and Biology of Cognitive Ethology.* Cambridge, Mass.: MIT Press.
Anderson, J. R. (1990) *The Adaptive Character of Thought.* Hillsdale, N.J.: Erlbaum.
Anderson, J. R. (1993) *Rules of the Mind.* Hillsdale, N.J.: Erlbaum.
Annett, J. (1996) On knowing how to do things: a theory of motor imagery. *Cognitive Brain Research* 3: 65–69.
Anscombe, G. E. M. ([1957] 1974) *Intention.* Reprint, Ithaca, N.Y.: Cornell University Press.
Aristotle (1985) *The Complete Works of Aristotle.* Vols. 1 and 2. Edited by J. Barnes. Princeton: Princeton University Press.
Aronson, E. ([1972] 1995) *The Social Animal.* Reprint, San Francisco: Freeman.
Asperger, H. (1944) Die autistischen Psychopathen im Kindesalter. *Archiv fur Psychiatrie under Nervenkrankheiten* 117: 76–136.
Astington, J. W. (1995) *The Child's Discovery of Mind.* Cambridge, Mass.: Harvard University Press.
Astington, J. W., and Gopnik, A. (1991) Developing understanding of desire and intention. In *Natural Theories of Mind,* edited by A. Whiten. Oxford: Basil Blackwell.
Astington, J. W., Harris, P. L., and Olson, D. R. (1988) *Developing Theories of Mind.* New York: Cambridge University Press.
Attwood, A., Frith U., and Hermelin, B. (1988) The understanding and use of interpersonal gestures by autistic and Down's syndrome children. *Journal of Autism and Developmental Disorders* 18: 241–257.
Aulkainen, S., Kulomaki, T., Hari, R. (1999) Normal movements reading in Asperger subjects. *Neuroreport* 10: 3467–3470.
Austin, J. L. (1970) *Philosophical Papers.* Oxford: Oxford University Press.
Austin, J. L. (1975) *How to Do Things with Words.* Cambridge, Mass.: Harvard University Press.
Avis, J., and Harris, P. L. (1991) Belief-desires reasoning among Baka children: Evidence for a universal conception of mind. *Child Development,* 62: 460–467.
Axelrood, J. (1984) *The Evolution of Cooperation.* New York: Basic.
Ayer, A. J. ([1936] 1952) *Language, Truth, and Language.* Reprint, New York: Dover.
Bachevalier, J. (1994) Medial temporal lobe structures and autism: A review of clinical and experimental findings. *Neuropsychologia* 6: 627–648.
Baldwin, D. (1991) Infants' ability to consult the speaker for clues to word reference. *Journal of Child Language* 20: 395–418.
Baltaxe, C. A. M. (1977) Pragmatic deficits in the language of autistic adolescents. *Journal of Pediatric Psychology* 2: 176–180.
Bamshad, M., Karom, M., Pallier, P., and Albers, H. E. (1997) Role of the central amygdala in social communication in Syrian hamsters *Mesocricetus auratus). Brain Research* 744: 15–22.
Bandura, A. (1986) *Social Foundations of Thought and Action.* Princeton, N.J.: Prentice-Hall.

Bard, K. A. (1992) Intentional behavior and intentional communication in young free-ranging orangutans. *Child Development* 63: 1186–1197.
Barkow, J. H., Cosmides, L., and Tooby, J. ([1992] 1995). *The Adapted Mind.* Reprint, Oxford: Oxford University Press.
Baron, J. (1997) *Thinking and Deciding.* 2nd ed. Cambridge: Cambridge University Press.
Baron-Cohen, S. (1989a) The autistic child's theory of mind: A case of specific developmental delay. *Journal of Child Psychology and Psychiatry* 30: 285–297.
Baron-Cohen, S. (1989b) Are autistic children behaviourists? An examination of their mental-physical and appearance-reality distinctions. *Journal of Autism and Developmental Disorders* 19: 579–600.
Baron-Cohen, S. (1989c) Joint attention deficits in autism: Towards a cognitive analysis. *Development and Psychopathology* 1: 185–189.
Baron-Cohen, S. (1995) *Mindblindness.* Cambridge, Mass.: MIT Press.
Baron-Cohen, S., and Gillberg, A. J. (1992) Can autism be detected at 18 months? The needle, the haystack, and the CHAT. *British Journal of Psychiatry* 161: 839–843.
Baron-Cohen, S., Jolliffe, L., Mortimore, C., and Robertson, M. (1997) Another advanced test of theory of mind: Evidence from very high functioning adults with autism and Asperger syndrome. *Journal of Child Psychology and Psychiatry* 38: 813–822.
Baron-Cohen, S., Leslie, A. M., and Frith, U. (1985) Does the autistic child have a "theory of mind"? *Cognition* 21: 37–46.
Baron-Cohen, S., Leslie, A. M., and Frith, U. (1986) Mechanical, behavioral, and intentional understanding of picture stories in autistic children. *British Journal of Developmental Psychology* 4: 113–125.
Baron-Cohen, S., Ring, H., Moriarty, J., Schmi, B., Costa, D., and Ell, P. (1994) Recognition of mental terms: Clinical findings in children with autism and a functional neuroimaging study of normal adults. *British Journal of Psychiatry* 165: 640–649.
Baron-Cohen, S., Ring, H. A., Wheelwright, S., Bullmore, E. T., Brammer, M. J., Simmons, A., and Williams, S. C. R. (1999) Social inteligence in the normal and autistic brain: An fMRI study. *European Journal of Neuroscience* 11: 1–8.
Baron-Cohen, S., Tager-Flusberg, H., and Cohen, D. J. ([1993] 1999) *Understanding Other Minds: Perspectives from Autism.* Reprint, Oxford: Oxford University Press.
Barresi, J., and Moore, C. (1996) Intentional relations and social understanding. *Behavioral and Brain Sciences* 19: 107–154.
Bartsch, K. (1996) Between desires and beliefs: Young children's action predictions. *Child Development* 67: 1671–1685.
Bartsch, K., and Estes, D. (1996) Individual differences in children's developing theory of mind and implications for metacognition. *Learning and Individual Differences* 8 (4): 281–304.

Bartsch, K., and Wellman, H. M. (1989) Young children's attribution of action to beliefs and desires. *Child Development* 60: 946–964.

Bartsch, K., and Wellman, H. M. (1995) *Children Talk about the Mind.* New York: Oxford University Press.

Bates, E. (1976) Pragmatics and sociolinguistics in child language. In *Normal and Deficient Child Language,* edited by D. M. Morehead and A. E. Morehead. Baltimore: University Park Press.

Bates, E. (1979) Intentions, conventions, and symbols. In *The Emergence of Symbols: Cognition and Communication in Infancy,* edited by E. Bates New York: Academic.

Bates, E., Benigni, L., Bretherton, I., Camaioni, L., and Volterra, V. (1979) *The Emergence of Symbols: Cognition communication in Infancy.* New York: Academic.

Bauman, M. L. (1991) Microscopic neuroanatomic abnormalaties in autism. *Pediatrics* 87: 791–896.

Bauman, M. L., and Kemper, T. L., (1994, 1997) *The Neurobiology of Autism.* Baltimore, MD: Johns Hopkins University Press.

Beach, F. A. (1942) Analysis of factors involved in the arousal, maintenance, and manifestation of sexual excitement in male animals. *Psychosomatic Medicine* 4: 173–198.

Bechara, A., Damasio, A. R., Damasio, H., and Anderson, S. W. (1994) Insensitivity to future consequences following damage to human prefrontal cortex. *Cognition* 50: 7–15.

Bechara, A., Damasio, H., Tranel, D., and Damasio, A. R. (1997) Deciding advantageously before knowing the advantagious strategy (see comments). *Science* 275 (5304): 1293–1295.

Bechara, A., Tranel, D., Damasio, H., and Damasio, A. R. (1996) Failure to respond autonomically to anticipated future outcomes following damage to prefrontal cortex. *Cerebral Cortex* 6: 215–225.

Bechtel, W. (1985) Realism, instrumentalism, and the intentional stance. *Cognitive Science* 9: 473–497.

Berger, P., and Luckmann, T. (1966) *The Social Construction of Reality.* New York: Anchor Doubleday.

Berlin, I. (1999) *The Roots of Romanticism.* Princeton: Princeton University Press.

Berridge, K. C. (1990) Comparative fine structure of action: Rules of form and sequence in the grooming patterns of six rodent species. *Behaviour* 113: 21–56.

Berridge, K. C. (1999) Pleasure, pain, desire, and dread: Hidden core processes of emotion. In *The Psychology of Well-Being,* edited by D. Kahneman, E. Diener, and N. Schwarz. New York: Russell Sage Foundation.

Berridge, K. C., Fentress, J. C., and Parr, H. (1987) Natural syntax rules control action sequence of rats. *Behavioral Brain Research* 23: 59–68.

Block, N. (1978) Troubles with functionalism. In *Perception and Cognition: Issues in the Foundations of Psychology,* edited by C. W. Savage. Minnesota Studies

in the Philosophy of Science, vol. 9. Minneapolis: University of Minnesota.
Block, N. (1980a) *Readings in the Philosophy of Psychology.* 2 vols. Cambridge, Mass.: Harvard University Press.
Block, N. (1980b) What is functionalism? In *Readings in Philosophy of Psychology,* edited by N. Block, vol. 1. Cambridge, Mass.: Harvard University Press.
Bloom, P. (1996) Intention, history, and artifact concepts. *Cognition* 60: 1–29.
Bloom, P., and Markson, L. (1998) Intention and analogy in children's naming of pictorial representations. *Psychological Science* 93: 200–204.
Boakes, R. (1984) *From Darwin to Behaviourism: Psychology and the Minds of Animals.* Cambridge: Cambridge University Press.
Boden, M. A. (1970) Intentionality and physical systems. *Philosophy of Science* 37: 200–214.
Boden, M. A. (1988) *Computer Models of Mind.* Cambridge: Cambridge University Press.
Bogdan, R. J. (1994) *Grounds for Cognition.* Hillsdale, N.J.: Lawrence Erlbaum.
Bogdan, R. J. (1997) *Interpreting Minds: The Evolution of a Practice.* Cambridge, Mass.: MIT Press.
Bookheimer, S. Y., Seffro, T. A., Blaxton, T., Gallard, W., and Theodore, W. (1995) Regional cerebral blood flow during object naming and word reading. *Human Brain Mapping* 3: 93–106.
Boring, E. G. (1952) The Gibsonian visual field. *Psychological Review* 59: 246–247.
Botterill, G. (1996) Folk, psychology, and theoretical status. In *Theories of Theories of Mind,* edited by P. Carruthers and P. K. Smith. Cambridge: Cambridge University Press.
Boucher, J. (1989) The theory of mind hypothesis of autism: Explanation, evidence, and assessment. *British Journal of Disorders of Communication* 24: 181–198.
Bourdieu, P. ([1972] 1991) *Outline of a Theory of Practice.* Translated by R. Nice. New York: Cambridge University Press.
Bourdieu, P. ([1980] 1990) *The Logic of Practice.* Translated by R. Nice. Stanford, Calif.: Stanford University Press.
Bower, G. H. (1993) The fragmentation of psychology? *American Psychologist* 8: 905–907.
Bower, T. G. R. (1979) *Human Development.* San Francisco: W. H. Freeman.
Bowlby, J. (1988) *A Secure Base: Parent-Child Attachment and Healthy Human Development.* New York: Basic.
Bowler, D. M. (1992) Theory of mind. I. Asperger's syndrome. *Journal of Child Psychology and Psychiatry* 33: 877–893.
Bradley, M. M., and Lang, P. L. (1999) Measuring emotion: Behavior, feeling, and physiology. In *Cognitive Neuroscience of Emotion,* edited by R. D. Lane and L. Nadel. Oxford: Oxford University Press.
Bratman, M. (1984) Two faces of intention. *Philosophical Review* 93: 375–405.

Bratman, M. (1987) *Intention, Plans, and Practical Reason.* Cambridge, Mass.: Harvard University Press.

Breitter, H. C., Etcoff, N. L., Whalen, P. J., Kennedy, W. A., Rauch, S. L., Buckner, R. L., Strauss, M. M., Hyman, S. E., and Rosen, B. R. (1996) Response and habituation of the human amygdala during visual processing of facial expression. *Neuron* 17: 875–887

Brentano, F. ([1874] 1995) *Psychology from an Empirical Standpoint.* Edited by O. Kraus, Translated by L. L. A. C. Rancurello, D. B. Terrell, and L. L. McAlister. London: Routledge and Kegan Paul.

Brentano, F. ([1929] 1981) *Sensory and Noetic Consciousness.* Edited by O. Kraus. Translated by M. Schattle and L. L. McAlister. New York: Humanities.

Bretherton, L., and Beeghly, M. (1982) Talking about internal states: The acquisition of an explicit theory of mind. *Developmental Psychology* 18: 906–921.

Brodal, A. ([1969] 1981) *Neurological Anatomy.* Reprint, Oxford: Oxford University Press.

Bronwell J., Griffin, R., Winner, E., Friedman, O., and Happe, F. (1999) Cerebral lateralization and theory of mind. In *Understanding Other Minds: Perspectives from Autism and Cognitive Neuroscience,* 2d ed., edited by S. Baron-Cohen, H. Tager-Flusberg, and D. Cohen. Oxford: Oxford University Press.

Brothers, L. (1990) The social brain: A project for integrating primate behavior and neurophysiology in a new domain. *Concepts in Neuroscience* 1: 27–51.

Brothers, L. (1994) Neurophysiology of the perception of intentions in primates. In *The Cognitive Neurosciences,* edited by M. S. Gazzaniga. Cambridge, Mass.: MIT Press.

Brothers, L. (1997) *Friday's Footprint.* Oxford: Oxford University Press.

Brothers, L., and Ring, B. (1993) Medial temporal neurons in the macaque monkey with responses selective for aspects of social stimuli. *Behavioral Brain Research* 57: 53–61.

Brothers, L., Ring, B., and Kling, A. (1990) Response of neurons in the macaque amygdala to complex social stimuli. *Behavioral Brain Research* 41: 199–213.

Brown, H. I. (1977) *Perception, Theory, and Commitment.* Chicago: University of Chicago Press.

Brown, P., and Marsden, C. D. (1998) What do the basal ganglia do? *Lancet* 351: 1901–1904.

Brunner, J. (1975a) From communication to language. *Cognition* 3: 255–287.

Brunner, J. (1975b) The ontogenesis of speech acts. *Journal of Child Language* 2: 1–20.

Brunner, J. (1983) *Child's Talk.* New York: Norton.

Bryant, P. E., and Trabasso, T. (1971) Transitive inferenes and memory in young children. *Nature* 232: 456–458.

Butterworth, G. (1991) The ontogeny and phylogeny of joint visual attention. In *Natural Theories of Mind: Evolution, Development, and Simulation of Everyday Mindreading,* edited by A. Whiten. Cambridge, Mass.: Blackwell.

Butterworth, G., and Cochran, E. (1980) Towards a mechanism of joint visual attention in human infancy. *International Journal of Behavioral Development* 3: 253–282.

Byrne, R. W. (1995) *The Thinking Ape*. Oxford: Oxford University Press.

Byrne, R. W., and Whiten, A. (1988) *Machiavellian Intelligence: Social Expertise and the Evolution of Intellect in Monkeys, Apes, and Humans*. New York: Oxford University Press.

Byrne, R. W., and Witten, R. (1992) Cognitive evolution in primates: Evidence from tactical deception. *Man* 27: 609–627.

Cadwaller, T. C. (1974) C. S. Peirce: The first American psychologist. *Journal of the History of the Behavioral Sciences* 10: 291–298.

Call, J., and Tomasello, M. (1998) Distinguishing intentional from accidental actions in oragutans *(Pongo pygmaes)*, chimpanzees *(Pan troglodytes)*, and human children *(Homo sapiens)*. *Journal of Comparative Psychology* 112: 192–206.

Calvert, G. A. Bullmore, E. T., Brammer, M. J., Campbell, R., Williams, S. C. R., McQuire, P. K., Woodruff, P. W. R., Iverson, S. D., and David, A. S. (1997) Activation of auditory cortex during silent lipreading. *Science* 276: 593–596.

Cannon, W. B. ([1915] 1963) *Bodily Changes in Pain, Hunger, Fear, and Rage*. Reprint, New York: Harper Torchbooks.

Caramazza, A., and Shelton, J. R. (1998) Domain-specific knowledge systems in the brain: The animate-inanimate distinction. *Journal of Cognitive Neuroscience* 10: 1–34.

Carey, S. (1985) *Conceptual Change in Childhood*. Cambridge, Mass.: MIT Press.

Carey, S., and Gelman, R. (1991) *The Epigenesis of Mind: Essays on Biology and Cognition*. Hillsdale, N.J.: Lawrence Erlbaum.

Carledge, B., ed. (1998) *Mind, Brain, and the Environment*. Oxford: Oxford University Press.

Carlson, S. M., Moses, L. J., and Hix, H. R. (1998) The role of inhibitory processes in young children's difficulties with deception and false belief. *Child Development* 69: 672–691.

Carnap, R. ([1947] 1975) *Meaning and Necessity: A Study in Semantics and Modal Logic*. Reprint, Chicago: University of Chicago Press.

Carnap, R. (1969) *The Logical Structure of the World and Pseudoproblems in Philosophy*. Translated by R. A. George. Berkeley and Los Angeles: University of California Press.

Carpenter, A. F., Georgopoulos, A. P., and Pellizzer, G. (1999) Motor cortical encoding of serial order in a context-recall task. *Science* 283: 1752–1757.

Carpenter, M. M., Nagell, K., and Tomasello, M. (1998) Social cognition, joint attention, and communicative competence from 9 to 15 months of age. *Monographs of the Society for Research in Child Development* 63: 1–133.

Carruthers, P. (1996) Simulation and self-knowledge: A defence of theory-theory. In *Theories of Theories of Mind,* edited by P. Carruthers and P. K. Smith. Cambridge: Cambridge University Press.

Carruthers, P., and Smith, P. K. (1996) *Theories of Theories of Mind.* Cambridge: Cambridge University Press.

Carter, C. S., Lederhendler, I. I., Kirkpatrick, B. ([1997] 1999) *The Integrative Neurobiology of Affiliation.* Reprint, Cambridge, Mass.: MIT Press.

Cassidy, K. W. (1998) Preschoolers' use of desire to solve theory of mind problems in a pretense context. *Developmental Psychology* 34: 503–511.

Chance, M. R. A., and Mead, A. P. (1953) Social behavior and primate evolution. *Symposia of the Society for Experimental Biology* 7: 395–439.

Chandler, M., and Helm, D. (1984) Developmental changes in the contributions of shared experience to social role taking competence. *International Journal of Behavioral Development* 7: 145–156.

Changeus, J. P., and Chavaillon J. (1995) *Origins of the Human Brain.* Oxford: Clarendon.

Chao, L. L., Haxby, J. V., and Martin, A. (1999) Attribute-based neural substrates in temporal cortex for perceiving and knowing objects. *Nature Neuroscience* 2: 913–919.

Charman, T., and Baron-Cohen, S. (1994) Another look at imitation in autism. *Development and Psychopathology* 6: 403–413.

Charman, T., Baron-Cohen, S., Swettenham, J., Cox, A., Baird, G., and Drew, A. (1997) Infants with autism: An investigation of empathy, pretend play, joint attention, and imitation. *Developmental Psychology* 33: 781–889.

Cheney, D. L., and Seyfarth, R. M. (1982a) How vervet monkeys perceive their grunts: Field play back experiments. *Animal Behaviour* 30: 739–851.

Cheney, D. L., and Seyfarth, R. M. (1982b) Recognition of individuals within and between groups of free-ranging vervet monkeys. *American Zoologist* 22: 519–529.

Cheney, D. L., and Seyfarth, R. M. (1985) Vervet monkey alarm calls: Manipulation through shared information? *Behaviour* 93: 150–166.

Cheney, D. L., and Seyfarth, R. M. (1986) The recognition of social alliances by vervet monkeys. *Animal Behaviour* 34: 1722–1731.

Cheney, D. L., and Seyfarth, R. M. (1988) Assessment of meaning and the detection of unreliable signals by vervet monkeys. *Animal Behaviour* 36: 477–486.

Cheney, D. L., and Seyfarth, R. M. (1989) Reconciliation and redirected aggression in vervet monkeys *(Cercopithecus aethiops). Behaviour* 110: 258–275.

Cheney D. L., and Seyfarth R. M. (1990a) Attending to behaviour versus attending to knowledge: Examining monkeys' attribution of mental states. *Animal Behaviour* 40: 742–853.

Cheney, D. L., and Seyfarth, R. M. (1990b) *How Monkeys See the World.* Chicago: University of Chicago Press.

Cheney, D. L., and Seyfarth, R. M. (1999) Function and intention in the calls of non-human primates. In *Evolution of Social Behavior Patterns in Primates and Man,* edited by W. G. Runclman, J. Maynard Smith, and R. I. M. Dunbar. Oxford: Oxford University Press.

Cheney, D. L., Seyfarth, R. M., and Silk, J. B. (1995) The responses of female baboons *(Papio cynocephalus ursinus)* to anomalous social interactions: Evidence for causal reasoning? *Journal of Comparative Psychology* 109: 134–141.

Chisholm, R. M., and Feehan, D. (1977) The intent to deceive. *Journal of Philosophy* 74 (3): 143–159.

Chomsky, N. ([1956] 1978) *Syntactic Structures.* Reprint, The Hague: Mouton.

Chomsky, N. (1959) Review of *Verbal Behavior* by B. F. Skinner. *Language* 35: 26–58.

Chomsky, N. (1965) *Aspects of the Theory of Syntax.* Cambridge, Mass.: MIT Press.

Chomsky, N. (1972) *Language and Mind.* New York: Harcourt, Brace and Jovanovich.

Christensen, S. M., and Turner, D. R. (1993) *Folk Psychology and the Philosophy of Mind.* Hillsdale, N.J.: Erlbaum.

Churchland, P. M. (1996) *The Engine of Reason, the Seat of the Soul: A Philosophical Journey into the Brain.* Cambridge, Mass.: MIT Press.

Churchland, P. M., and Churchland, P. S. (1998) *On the Contrary.* Cambridge, Mass.: MIT Press.

Churchland, P. S. (1986) *Neurophilosophy.* Cambridge, Mass.: MIT Press.

Churchland, P. S. (1992) *The Computational Brain.* Cambridge, Mass.: MIT Press.

Clark, A. ([1996] 1997) *Being There.* Reprint, Cambridge, Mass.: MIT Press.

Clark, R. E., and Squire, L. R. (1998) Classical conditioning and brain systems: The role of awareness. *Science* 280: 77–81.

Cohen, M. S., Kooslyn, S. M., Breitter, H. C., DiGirolamo, G. J., Thompson, W. L., Anderson, A. K., Bookheimer, S. Y., Rosen, B. R., and Belliveau, J. W. (1996) Changes in cortical activity during mental rotation: A mapping study using functional MRI. *Brain* 119: 89–100.

Cole, J. (1998) *About Face.* Cambridge, Mass.: MIT Press.

Coleman, M., and Gillberg, C. (1985) *The Biology of the Autistic Syndromes.* New York: Praeger.

Cook, E. H., and Leventhal, B. L. (1996) The serotonin system in autism. *Current Opinion in Pediatrics* 8: 348–354.

Corballis, M. C., and Lea, S. E. G. (1999) *The Descent of Mind.* Oxford: Oxford University Press.

Corona, R., Dissanayake, C., Arbelle, S., Wellington, P., and Sigman, M. (1998) Is affect aversive to young children with autism? Behavioral and cardiac responses to experimenter distress. *Child Development* 69: 1494–1502.

Cosmides, L. (1989) The logic of social exchange: Has natural selection shaped how humans reason? Studies with the Wason selection task. *Cognition* 31: 187–276.

Cosmides, L., and Tooby, J. (1994) Beyond intuition and instinct blindness: Toward an evolutionarily rigorous cognitive science. *Cognition* 50: 41–87.

Courchesne, E. et al. (1994) Impairment in shifting attention in autistic and cerebellar patients. *Behavioral Neuroscience* 108: 848–865.

Craig, W. C. (1918) Appetites and aversions as constituents of instincts. *Biological Bulletin* 34: 91–107.

Curcio, F. (1978) Sensorimotor functioning and communication in mute autistic children. *Journal of Autism and Childhood Schizophrenia* 8: 281–292.

Damasio, A. R. (1994) *Descartes' Error: Emotion, Reason, and the Human Brain.* New York: Putnam.

Damasio, A. R. (1996) The somatic marker hypothesis and the possible functions of the prefrontal cortex. *Proceedings of the Royal Society of London* 351: 1413–1420.

Damasio, A. R., and Maurer, R. G. (1978) A neurological model for childhood autism. *Archives of Neurology* 35: 777–786.

Damasio, A. R. Tranel, D., and Damasio, H. (1990) Individuals with sociopathic behavior caused by frontal damage fail to respond automatically to social stimuli. *Behavioural Brain Research* 41: 81–94.

Damasio, H. et al. (1996) A neural basis for lexical retrieval. *Nature* 380: 499–505.

Darwin, C. ([1859] 1958) *The Origin of Species.* Reprint, New York: Mentor.

Darwin, C. ([1872] 1965) *The Expression of the Emotions in Man and Animals.* Reprint, Chicago: University of Chicago Press.

Dasser, V., Ulbaek, I., and Premack, D. (1989) The perception of intention. *Science* 243 (4889): 365–367.

Davidson, D. (1963) Actions, reasons, and causes. *Journal of Philosophy* 60: 685–800.

Davidson, D. (1980) *Essays on Actions and Events.* Oxford: Clarendon.

Davidson, R. J. (1992) Anterior cerebral asymmetry and the nature of emotion. *Brain and Cognition* 20: 125–151.

Davidson, R. J., and Rickman, M. (1999) Behavioral inhibition and the emotional circuitry of the brain: Stability and plasticity during the early childhood years. In *Extreme Fear, Shyness, and Social Phobia,* edited by L. A. Schmidt and J. Schulkin. Oxford: Oxford University Press.

Davidson, R. J., and Sutton, S. K. (1995) Affective neuroscience: The emergence of a discipline. *Current Opinion in Neurobiology* 5: 217–224.

Dawkins, R. ([1976] 1989) *The Selfish Gene.* New edition. New York: Oxford University Press.

Dawkins, R. (1982) *The Extended Phenotype.* New York: Oxford University Press.

Dawkins, R., and Krebs, J. R. (1978) Animal signals: Information or manipulations? In *Behavioral Ecology: An Evolutionary Approach,* edited by J. R. Krebs and N. B. Davies. Oxford: Blackwell Scientific Publications.

Dawson, G., and Adams, A. (1984) Imitation and social responsiveness in autistic children. *Journal of Abnormal Child Psychology* 12: 209–226.

Decety, J. (1996) Do imagined and executed actions share the same neural substrate? *Cognitive Brain Research* 3: 87–93.
Decety, J., Durozard, D., and Baverel, G. (1993) Central activation of autonomic effectors during mental simulation of motor actions in man. *Journal of Physiology* 461: 549–563.
Decety, J., Grezes, J., Costes, N., Perani, D., Jeannerod, M., Procyk, E., Grassi, F., and Fazio, F. (1997) Brain activity during observation of actions: Influence of action content and subject's strategy. *Brain* 120: 1763–1777.
Decety, J., Perani, D., and Jeannerod, M. (1994) Mapping motor representations with positron emission tomography. *Nature* 371: 600–602.
Dennett, D. C. (1969) *Content and Consciousness.* London: Routledge and Kegan; New York: Humanities.
Dennett, D. C. (1978) *Brainstorms: Philosophical Essays on Mind and Psychology.* Brighton: Harvester.
Dennett, D. C. (1987) *The Intentional Stance.* Cambridge, Mass.: MIT Press.
Dennett, D. C. (1991) *Consciousness Explained.* Boston: Little, Brown.
Dennett, D. C (1996) *Kinds of Minds.* New York: Basic.
Descartes, René ([1628] 1970) Rules for the direction of the understanding. In *The Philosophical Works of Descartes,* 2 vols., edited by E. S. Haldane and G. R. T. Ross. Cambridge: Cambridge University Press.
Desimone, R. (1991) Face-selective cells in the temporal cortex of monkeys. *Journal of Cognitive Neuroscience* 3: 1–8
Desimone, R., Albright, T. D., Gross, C. G., and Bruce, C. (1984) Stimulus-selective properties of inferior temporal neurons in the macaque. *Journal of Neuroscience* 8: 2051–2062.
De Sousa, R. ([1987] 1995) *The Rationality of Emotion.* Reprint, Cambridge, Mass.: MIT Press.
DeWaal, F. B. M. (1997) *Bonobo, the Forgotten Ape.* Berkeley and Los Angeles: University of California Press.
DeWaal, F. B. M., and Lanting, G. (1989) *Peacemaking among Primates.* Cambridge, Mass.: Harvard University Press.
DeWaal, F. B. M., van Hooff, J. A. R. A. M., and Netto, W. J. (1976) An ethological analysis of types of agonistic interaction in a captive group of Java-monkeys *(Macaca fascicularis). Primates* 17: 257–290.
Dewey, J. (1894) The theory of emotions. I. Emotional attitudes. *Psychological Review* 1: 553–569.
Dewey, J. (1895) The theory of the emotions. II. The significance of the emotions. *Psychological Review* 2: 13–32.
Dewey, J. (1896) The reflex arc concept in psychology. *Psychological Review* 111: 357–370.
Dewey, J. (1910) *How We Think.* New York: D. C. Heath.
Dewey, J. ([1910] 1965) *The Influence of Darwin on Philosophy.* Reprint, Bloomington: Indiana University Press.
Dewey, J. (1916) *Essays in Experimental Logic.* New York: Dover.

Dewey, J. ([1925] 1989) *Experience and Nature.* Reprint. LaSalle, Ill.: Open Court Press.

Dewsbury, D. A. (1991) Psychobiology. *American Psychologist* 46: 198–205.

Dickinson, A. (1980) *Contemporary Animal Learning Theory.* Cambridge: Cambridge University Press.

Dickinson, A., and Shanks, D. (1995) Instrumental action and causal representation. In *Causal Cognition,* edited by D. Sperber, D. Premack, and A. J. Premack. Oxford: Clarendon.

Doi, T. (1973) *The Anatomy of Dependence.* Tokyo; Kodansha.

Dolgin, K. G., and Behrend, D. A. (1984) Children's knowledge about animates and inanimates. *Child Development* 55: 1646–1650.

Dominey, P., Devety, J., Brousolle, E., Chazot, G., and Jeannerod. M. (1995) Motor imagery of a lateralized sequential task is assymetrically slowed in hemi-Parkinson patients. *Neuropsychologia* 33: 727–841.

Dreecke, L. (1996) Planning, preparation, execution, and imagery of volitional action. *Cognitive Brain Research* 3: 59–64.

Dretske, F. I. (1981) *Knowledge and the Flow of Information.* Cambridge, Mass.: MIT Press.

Dretske, F. I. (1988) *Explaining Behavior: Reasons in a World of Causes.* Cambridge, Mass.: MIT Press.

Dretske, F. I. (1995) *Naturalizing the Mind.* Cambridge, Mass.: MIT Press.

Dreyfus, H., and Dreyfus, S. (1988) Making a mind versus modeling the brain: Artificial intelligence back at a branchpoint. *Daedalus* (winter): 15–43.

Duchenne, B. ([1862] 1990). *The Mechanism of Human Facial Expression or an Electro-physiological Analysis of the Expression of the Emotions.* Translated by A. Cuthbertson. New York: Cambridge University Press.

Dunbar, R. I. M. (1988) *Primate Social Systems.* London: Croom Helm.

Dunbar, R. I. M. (1995) Neocortex size and group size in primates: A test of the hypothesis. *Journal of Human Evolution* 28: 287–296.

Dunn, J. (1988) *The Beginnings of Social Understanding.* Oxford: Blackwell.

Eco, U. (1976) *A Theory of Semiotics.* Bloomington: Indiana University Press.

Eddy, T. J., Gallup, G. G., Jr., and Povinelli, D. J. (1996) Age differences in the ability of chimpanzees to distinguish mirror-images of self from video images of others. *Journal of Comparative Psychology* 1101: 38–44.

Ekman, P. (1973) Cross-cultural studies of facial expression. In *Darwin and Facial Expression,* edited by P. Ekman. New York: Academic.

Ekman, P. (1994) Strong evidence for universals in facial expressions: A reply to Russell's mistaken critique. *Psychological Bulletin* 115: 268–287.

Ekman, P., and Davidson, R. J. (1994) *The Nature of Emotion.* New York: Oxford University Press.

Ekman, P., Sorenson, E. R., and Friesen, W. V. (1969). Pan-cultural elements in facial displays of emotions. *Science* 164: 86–88.

Eldredge, N. (1985) *Unfinished Synthesis, Biological Hierarchies, and Modern Evolutionary Thought.* New York: Oxford University Press.

Elster, J. ([1979] 1988) *Ulysses and the Sirens.* Reprint, Cambridge: Cambridge University Press.

Emery, N. J., and Amaral, D. G. (1999) The role of the amygdala in primate social cognition. In *Cognitive Neuroscience of Emotion,* edited by R. D. Lane and L. Nadel. Oxford: Oxford University Press.

Ericsson, K. A., and Simin, H. A. (1980) Verbal reports as data. *Psychological Review* 87: 215–251.

Estes, D. (1998) Young children's awareness of their mental activity: The case of mental rotation. *Child Development* 69: 1345–1360.

Everitt, B. J., Cador, M., and Robbin, T. W. (1989) Interactions between the amygdala and ventral striatum in stimulus-reward associations: Studies using a second-order schedule of sexual reinforcement. *Neuroscience* 30: 63–85.

Farah, M. J. (1984) The neurobiological basis of visual imagery: A componential analysis. *Cognition* 18: 245–272.

Farah, M. J. (1990) *Visual Agnosia.* Cambridge, Mass.: MIT Press.

Fein, D. A. (1972) Judgments of causality to physical and social sequences. *Developmental Psychology* 8: 147–153.

Fein, D., Penington, B., Markovitz, P., Braverman, M., and Waterhouse, L. (1986) Toward a neuropsychological model of infantile autism: Are the social deficits primary? *Journal of American Academic Childhood Psychiatry* 25: 198–212.

Feldman, C. F. (1992) The new theory of theory of mind. *Human Development* 35: 107–117.

Felleman, D. J., and Van Essen, D. C. (1991). Distributed hierarchial processing in the primate cerebral cortex. *Cerebral Cortex* 1: 1–47.

Fink, G. R., Marshall, J. C., Halligan, P. W., Frith, C. D., Driver, J., Frackowiak, R. S. J., and Dolan, R. J. (1999) The neural consequences of conflict between intention and the senses. *Brain* 122: 497–512.

Fiske, A. P. (1991) *Structures of Social Life: The Four Elementary Forms of Human Relations.* New York: Free Press.

Fiske, S. (1993) Social cognition and social perception. *Annual Review of Psychology* 44: 155–194.

Fiske, S., and Taylor, S. E. (1991) *Social Cognition.* New York: McGraw-Hill.

Flanagan, O. (1996) *Self Expression.* New York: Oxford University Press.

Flavell, J. H. (1986) The development of children's knowledge about the appearace-reality distinction. *American Psychologist* 41: 418–425.

Flavell, J. H. (1988) The development of children's knowledge about the mind. In *Developing Theories of Mind,* edited by J. W. Astington, P. L. Harris, and D. R. Olson. Cambridge: Cambridge University Press.

Flavell, J. H. (1999) Cognitive development: Children's knowledge about the mind. *Annual Review of Psychology* 50: 21–45.

Flavell, J. H., Green, F. L., and Flavell, E. R. (1995) Young children's knowledge about thinking. *Monograph Society Research Child Development* 60: 1–96, discussion 97–114.

Flavell, J. H., and Miller, P. H. (1998) Social cognition. In *Handbook of Child Psychology,* edited by D. Kuhn and R. S. Siegler. New York: John Wiley.

Fletcher, P. C., Happe, F., Frith, U., Baker, S. C., Dolan, R. J., Frackowiak, R. S., and Frith, C. D. (1991) Other minds in the brain: A functional imaging study of "theory of mind" in story comprehension. *Cognition* 57: 109–128.

Fodor, J. A. (1981) *Representations.* Cambridge, Mass.: MIT Press.

Fodor, J. A. (1983) *The Modularity of Mind.* Cambridge, Mass.: MIT Press.

Fodor, J. A. (1987) *Psychosemantics.* Cambridge, Mass.: MIT Press.

Fodor, J. A. (1990) *A Theory of Content and Other Essays.* Cambridge, Mass.: MIT Press.

Fodor, J. A. (1992) A theory of the child's theory of mind. *Cognition* 44: 283–296.

Fodor, J. A. ([1994] 1995) *The Elm and the Expert.* Reprint, Cambridge, Mass.: MIT Press.

Fodor, J. A., (1998) *Concepts.* Oxford: Oxford University Press.

Folstein, S., and Rutter, M. (1977) Infantile autism: A genetic study of 21 twin pairs. *Journal of Child Psychology and Psychiatry* 18: 297–321.

Fonberg, E. (1974) Amygdala functions within the alimentary system. *Acta Neurobiologiae Experimentalis* 22: 51–57.

Fonberg, E., and Kostarczyk, E. (1980) Motivational role of social reinforcement in dog-man relations. *Acta Neurbiologiae Experimentalis* 40: 117–136.

Foucault, M. (1971) *The Order of Things: An Archeology of the Human Science.* New York: Pantheon.

Fox, N. A., and Davidson, R. J. (1988) Patterns of brain electrical activity during facial signs of emotion in 10-month-old infants. *Developmental Psychology* 24: 230–236.

Fox, N., Rubin, K. H., Calkins, S. D., Marshall, T. R., Coplan, R. J., Porges, S. W., Long, J. M., and Steward, S. (1995) Frontal activation and social competence at four years of age. *Child Development* 66: 1770–1784.

Frege, G. ([1892] 1966) On sense and reference. In *Translations from the Writings of Gottlob Frege,* 2d ed., edited by P. Geach and M. Black. Oxford: Blackwell.

Frege, G. (1977) *Logical Investigations.* Edited by P. T. Geach. Translated by P. T. Geach and R. H. Stoothoff. New Haven, Conn.: Yale University Press.

Fridlund, A. (1991) Evolution and facial action in reflex, social motive, and paralanguage. *Biological Psychology* 32: 3–100.

Friedman, M. (1992) *Kant and the Exact Sciences.* Cambridge, Mass.: Harvard University Press.

Frijda, N. (1986) *The Emotions.* Cambridge: Cambridge University Press.

Frith, C. D., and Frith, U. (1999) Interacting minds—biological basis. *Science* 286: 1692–1695.

Frith, U. (1972) Cognitive mechanisms in autism: Experiments with color and tone sequence production. *Journal of Autism and Childhood Schizophrenia* 2: 160–183.

Frith, U. (1989) *Autism: Explaining the Enigma.* Cambridge, Mass.: Basil Blackwell.
Frith, U. ([1991] 1997) *Autism and Asperger Syndrome.* Reprint, Cambridge, Cambridge University Press.
Frith, U. (1993) Autism. *Scientific American* (June).
Frith, U. (1995) Autism: Beyond "theory of mind." *Cognition* 50: 13–30.
Frith, U., and Hermelin, G. (1969) The role of visual and motor cues for normal, subnormal, and autistic children. *Journal of Child Psychology and Psychiatry* 10: 13–63.
Frith, U., Morton, J., and Leslie, A. M. (1991) The cognitive basis of a biological disorder: Autism. *TINS* 14: 433–438.
Frye, D., and Moore, C. (1991) *Children's Theories of Mind: Mental States and Social Understanding.* Hillsdale, N.J.: Lawrence Erlbaum.
Frye, D., Zelazo, P. D., and Burack, J. A. (1998) Cognitive complexity and control: 1. Theory of mind in typical and atypical development. *Current Directions in Psychological Science* 7: 116–117.
Funnell, E., and Sheridan, J. (1992) Categories of knowledge? Unfamiliar aspects of living and non-living things. *Cognitive Neuropsychology* 9: 135–154.
Gadamer, H. G. (1975) *Truth and Method.* New York: Crossroad.
Gadamer, H. G. (1982) Reason in the Age of Science. Cambridge, Mass.: MIT Press.
Galef, B. G. (1990) Tradition in animals: Field observations and laboratory analysis. In *Methods, Inferences, Interpretation, and Explanation in the Study of Behavior,* edited by M. Bekoff and D. Jamieson. Boulder, Colo.: Westview.
Gallagher, M., and Holland P. C. (1994) The amygdala complex: Multipyle roles in associative learning and attention. *Proceedings of the National Academy of Sciences* 91: 11771–11776.
Gallagher, H. L., Happe, F., Brunswick, N., Fletcher, P. C., Frith, U., and Frith, C. D. (2000) Reading the mind in cartoons and stories: An fMRI study of "theory of mind" in verbal and nonverbal tasks. *Neuropsychologia* 38: 11–21.
Gallese, V., Fadiga, L., Fogassi, L., and Rizzolatti, G. (1996) Action recognition in the premotor cortex. *Brain* 119: 593–609.
Gallese, V., and Goldman, A. (1999) Mirror neurons and the simulation theory of mind reading. *Cognitive Science* 12: 493–501.
Gallistel C. R. (1990) *The Organization of Learning.* Cambridge, Mass.: MIT Press.
Gallup, G., Jr. (1970) Chimpanzees: Self-recognition. *Science* 167: 86–87.
Gallup, G., Jr. (1982) Self-awareness and the emergence of mind in primates. *American Journal of Primatology* 2: 237–248.
Gallup, G. G., Jr., Povinelli, D. J., Suarez, S. D., Anderson, J. R., Lethmate, J., and Menzel, E. W. (1995) Further reflections on self-recognition in primates. *Animal Behavior* 50: 1525–1532.

Garcia, J., and Ervin, F. R (1968) Gustatory-visceral and telereceptor-cutaneous conditioning-adaptation to internal and external milieus. *Communications in Behavioral Biology* A1: 389–415.

Garcia, J., Hankins, W. G., and Rusinak, K. W. (1974) Behavioral regulation of the internal milieu in man and rat. *Science* 185: 824–831.

Gardner, B. T., and Gardner, R. A. (1971) Two-way communication with an infant chimpanzee. In *Behavior of Non-Human Primates*, vol. 4, edited by A. Schrier and F. Stollnitz. New York: Academic.

Gardner, H. (1985) *The Mind's New Science.* New York: Basic.

Garfinkel, A. (1981) *Forms of Explanation.* New Haven, Conn.: Yale University Press.

Gazzaniga, M. S. (1985) *The Social Brain: Discovering the Networks of the Mind.* New York: Basic.

Gazzaniga, M. S., ed. (1995) *The Cognitive Neurosciences.* Cambridge, Mass.: MIT Press.

Gazzaniga, M. S., Bogen, J. E., and Sperry, R. W. (1962) Some functional effects of sectioning the cerebral commissures in man. *National Academy of Sciences* 48: 1765–1769.

Geach, P. T. (1957) *Mental Acts: Their Content and Their Objects. Studies in Philosophical Psychology.* London: Routledge and Kegan Paul.

Geertz, C. (1973) *The Interpretation of Culture.* New York: Basic.

Gelman, R. A., and Gallistel, C. R. (1978) *The Child's Understanding of Number.* Cambridge, Mass.: Harvard University Press.

Gelman, R., and Spelke, E. (1981) The development of thoughts about animate and inanimate objects: Implications for research on social cognition. In *Social Cognitive Development*, edited by J. H. Flavell and L. Ross. Cambridge: Cambridge University Press.

Gelman, S. A., and Coley, J. D. (1990) The importance of knowing a dodo is a bird: Categories and inferences in 2-year-old children. *Developmental Psychology* 26: 796–804.

Gelman, S. A., and Markman, E. (1986) Categories and induction in young children. *Cognition* 23: 183–209.

Gelman, S. A., and Wellman, H. M. (1991) Insides and essence: Early understanding of the non-obvious. *Cognition* 38: 213–244.

Gergen, K. (1991) *The Saturated Self: Dilemmas of Identity in Contemporary Life.* New York: Basic.

Geschwind, N. (1974) *Selected Papers on Language and the Brain.* Boston Studies in the Philosophy of Science, vol. 16, edited by R. S. Cohen and M. W. Wartofsky. Dordrecht-Holland: D. Reidel Publ. Co.

Geschwind, N. (1980) Neurological knowledge and complex behaviors. *Cognitive Science* 4: 185–193.

Geschwind N. (1981) Neurological knowledge and complex behaviors. In *Perspectives on Cognitive Science*, edited by Donald A. Norman. Hillsdale, N.J.: Ablex, Lawrence Erlbaum.

Gibson, E., and Rader, N. (1979) Attention: The perceiver as performer. In *Attention and Cognitive Development,* edited by G. A. Hale and M. Lewis. New York: Plenum.

Gibson, J. J. (1966) *The Senses Considered as Perceptual Systems.* Boston, Mass.: Houghton Mifflin.

Gibson J. J. (1979) *The Ecological Approach to Visual Perception.* New York: Houghton Mifflin.

Gibson, J. J. (1982) Reasons for realism. In *Selected Essays of J. J. Gibson,* edited by E. Reed and R. Jones. Montclair, N.J.: Erlbaum.

Glanzer, M. (1968) Storage mechanisms in free recall. *Transactions of the New York Academy of Sciences* 30: 1120–1129.

Glymour, C. (1992) *Thinking Things Through: An Introduction to Philosophical Issues and Achievements.* Cambridge, Mass.: MIT Press.

Goel, V., Grafman, J., Sadato, N., and Hallett, M. (1995) Modeling other minds. *Neuroreport* 6: 1741–1746.

Goel, V., Grafman, J., Tajik, J., Gana, S., and Danto, D. (1997) A study of the performance of patients with frontal lobe lesions in a financial planning task. *Brain* 10: 1805–1822.

Goffman, E. (1959) *The Presentation of Self in Everyday Life.* New York: Doubleday Anchor.

Goffman, E. (1971) *Relations in Public.* New York: Harper and Row.

Goffman, E. (1981) *Forms of Talk.* Oxford: Blackwell.

Goldman, A. (1993) The psychology of folk psychology. *Mind and Language* 7 (10): 4–19.

Goldman-Rakic, P. S. (1996) The prefrontal landscape: Implications of functional architecture for understanding human mentation and the central executive. *Philosophical Transactions of the Royal Society of London—Series B* 351: 1445–1453.

Gomez, J. C. (1990) The emergence of intentional communication as a problem-solving strategy in the gorilla. In *"Language" and Intelligence in Monkeys and Apes,* edited by S. T. Parker and K. R. Gibson. Cambridge: Cambridge University Press.

Gomez, J. C. (1991) Visual behavior as a window for reading the mind of others in primates. In *Natural Theories of Mind,* edited by A. Whiten. Oxford: Blackwell.

Gomez, J. C. (1996) Non-human primate theories of mind. In *Theories of Theories of Mind,* edited by P. Carruthers and P. K. Smith. Cambridge: Cambridge University Press.

Gomez, J. C., Sarria, E., and Tamarit, J. (1993) The comparative study of early communication and theories of mind: Ontogeny, Phylogeny, and pathology. In *Understanding Other Minds: Perspectives from Autism,* edited by S. Baron-Cohen, H. Tager-Flusberg, and D. J. Cohen. Oxford: Oxford University Press.

Goodall, J. (1986) *The Chimpanzees of Gombe: Patterns of Behavior.* Cambridge, Mass.: Harvard University Press.

Goodman, N. (1972) *Problems and Projects.* Indianapolis: Bobbs-Merrill.

Gopnik, A. (1988) Conceptual and semantic development as theory change. *Mind and Language* 3: 197–217.

Gopnik, A. (1990) Developing the idea of intentionality: Children's theories of mind. *Canadian Journal of Philosophy* 20: 89–113.

Gopnik, A. (1993) How we know our minds. *Behavioral and Brain Sciences* 16: 1–14.

Gopnik, A., and Astington, J. W. (1988) Children's understanding of representational change and its relation to the understanding of false belief and the appearance-reality distinction. *Child Development* 59: 26–37.

Gopnik, A., and Graf, P. (1988) Knowing how you know: Young children's ability to identify and remember the sources of their beliefs. *Child Development* 59: 1366–1371.

Gopnik, A., and Meltzoff, A. N. (1997) *Words, Thoughts, and Theories.* Cambridge, Mass.: MIT Press.

Gopnik, A., and Slaughter, V. (1991) Young children's understanding of changes in their mental states. *Child Development* 62: 98–110.

Gopnik, A., and Wellman, H. M. (1992) Why the child's theory of mind really is a theory. *Mind and Language* 7: 145–171.

Gray, T. S., Magnuson, D. (1987) Neuropeptide neuronal efferents from the bed nucleus of the stria terminalis and central amygdaloid nucleus to the dorsal vagal compex in the rat. *Journal of Comparative Neurology* 262: 365–374.

Graybiel, A. M. (1995) Building action repertoires: Memory and learning functions of the basal ganglia. *Current Opinion in Neurobiology* 5: 733–841.

Graybiel, A. M. (1997) The basal ganglia and cognitive pattern generators. *Schizophrenia Research* 23: 459–469.

Graybiel, A. M. (1998) The basal ganglia and chunking of action repertoires. *Neurobiology of Learning and Memory* 70: 119–136.

Grice, H. P. (1957) Meaning. *Philosophical Review* 66: 377–388.

Grice, H. P. (1975) Logic and Conversation. In *Syntax and Semantics,* vol. 3, *Speech Acts,* edited by P. Cole and J. Morgan. New York: Academic Press.

Grice, H. P. (1989) *Studies in the Way of Words.* Cambridge, Mass.: Harvard University Press.

Griffin, D. R. (1958) *Listening in the Dark.* New Haven, Conn.: Yale University Press.

Gross, C. G. (1992). Representation of visual stimuli in inferior temporal cortex. *Philosophical Transactions of the Royal Society of London—Series B, Biological Sciences* 335: 3–10.

Gunnar, M. R., and Maratsos, M. (1992) *Modularity and Constraints in Language and Cognition,* vol. 25. Hillsdale, N.J.: Lawrence Erlbaum.

Gyger, M., Karakashian, S. J., Duffy, A. M., Jr., and Marler, P. (1988) Alarm signals in birds: The role of testosterone. *Hormones and Behavior* 22: 305–314.

Habermas J. ([1990] 1991) *Moral Consciousness and Communicative Action.* Translated by C. Lenhardt and S. W. Nicholsen. Cambridge, Mass.: MIT Press.

Hamlyn, D. W. (1978) *Experience and the Growth of Understanding.* London: Routledge and Kegan Paul.

Hanson, N. R. (1958) *Patterns of Discovery.* Cambridge and New York: Cambridge University Press.

Happé, F. (1994) Reading minds through a glass darkly. *Cahiers de Psychologie Cognition* 13: 599–606.

Happé, F., Bronell, H., and Winner, E. (1999) Acquired theory of mind impairments following right hemispheric stroke. *Cognition* 70: 211–240.

Happé, F., and Frith, U. (1996) The neuropsychology of autism. *Brain* 119: 1377–1400.

Happé, F., Ehlers, S., Fletcher, P., Frith, U., Johansson, M., Gillberg, C., Dolan, R., Frackowiak, R., and Frith, C., (1996) "Theory of Mind" in the brain. Evidence from a PET scan study of Asperger syndrome. *Neuroreport* 8: 197–201.

Happé, F., Winner, E., and Bronell, H. (1998) The getting of wisdom: Theory of mind in old age. *Developmental Psychology* 36: 858–862.

Haraway, D. (1989) *Primate Visions.* London: Routledge.

Harding, C. G., and Golinkoff, R. M. (1979) The origins of intentional vocalizations in prelinguistic infants. *Child Development* 50: 33–40.

Hari, R., Forss, N., Avikainen, S., Kirverskari, E., Salenius, S., and Rizzolatti, G. (1998) Activation of human primary cortex during action observation: A neuromagnetic study. *Proceedings of the National Academy of Sciences* 15061–15065.

Harnard, S., ed. ([1987] 1990) *Categorical Perception.* Reprint, Cambridge, Mass.: MIT Press.

Harries, M. H., and Perrett, D. I. (1991). Modular organization of face processing in temporal cortes: Physiological evidence and possible anatomical correlates. *Journal of Cognitive Neuroscience* 3: 9–24.

Harris, N. S., Courchesne, E., Townsend, J., Carper, R. A., and Lord, C. (1999). Neuroanatomic contributions to slowed orienting of attention in children with autism. *Cognitive Brain Research* 8: 61–81.

Harris, P. L. (1989) The autistic child's impaired conception of mental states. *Development and Psychopathology* 1: 191–196.

Harris, P. L. (1989) *Children and Emotion.* Oxford: Blackwell.

Harris, P. L. (1996) Desires, beliefs, and language. In *Theories of Theories of Mind,* edited by P. Carruthers and P. K. Smith. Cambridge: Cambridge University Press.

Haugeland, J., ed. ([1981] 1991) *Mind Design: Philosophy, Psychology, Artificial Intelligence.* Reprint, Cambridge, Mass.: MIT Press.

Haugeland, J. (1998) *Having Thought.* Cambridge, Mass.: Harvard University Press.

Hauser, M. D. (1996) *The Evolution of Communication.* Cambridge, Mass.: MIT Press.

Hauser, M. D. (1999) Perserveration, inhibition, and the prefrontal cortex: A new look. *Current Opinion in Neurobiology* 9: 214–222.

Hauser, M. D., and Carey, S. (1998) Building a cognitive creature from a set of primitives: evolutionary and developmental insights. In *The Evolution of Mind,* edited by D. D. Cummins and C. Allen.

Hauser, M. D., Kralik, J., Botto-Mahan, C., Garrett, M., and Oser, J. (1995) Self-recognition in primates: Phylogeny and the salience of species-typical features. *Proceedings of the National Academy of Sciences* (USA) 92: 10811–10814.

Hauser, M. D., and Marler P. (1993) Food-associated calls in rhesus macaques *(Macaca mulatta):* 1. Socioecological factors. *Behavioral Ecology* 4: 194–205.

Haxby, J. V., Ungerleider, L. G., Clark, V. P., Schouten, J. L., Hoffman, E. A., and Martin, A. (1999) The effect of face inversion on activity in human neural systems for face and object perception. *Neuron* 22: 189–199.

Hayes, K. J., and Hayes, C. H. (1951) The intellectual development of a home-raised chimpanzee. *Proceedings of the American Philosophical Society* 95: 105–109.

Hayes, K. J., and Hayes, C. H. (1971) Higher mental functions of a home-raised chimpanzee. In *Behavior of Nonhuman Primates,* edited by A. M. Schrier and F. Stolntz. New York: Academic.

Hebb, D. O. (1949) *The Organization of Behavior.* New York: Wiley.

Heelan, P. A. (1983) *Space Perception and the Philosophy of Science.* Berkeley and Los Angeles: University of California Press.

Heelan, P. A., and Schulkin, J. (1998). Hermeneutical philosophy and pragmatism: A philosophy of science. *Synthese* 115: 269–302.

Heidegger, M. ([1927] 1962) *Being and Time.* Translated by J. Macquarrie and E. Robinson. New York: Harper and Row.

Heil, J., and Mele, A. (1995) *Mental Causation.* Oxford: Oxford University Press.

Helmholtz, H. (1873) *Popular Lectures in Scientific Subjects.* Translated by E. Atkinson. Introduction by J. Tyndale. London: Longmans Green.

Hempel, C. G. ([1965] 1970) *Aspects of Scientific Explanation and Other Essays in the Philosophy of Science.* Reprint, New York: Free Press.

Herbert, J. (1993) Peptides in the limbic system: Neurochemical codes for coordinated adaptive responses to behavioral and physiological demand. *Neurobiology* 41: 723–891.

Herrick, C. J. (1905) The central gustatory pathway in the brain of bony fishes. *Journal of Comparative Neurology* 15: 375–456.

Herrick, C. J. (1926) *Brains in Rats and Men.* Chicago: University of Chicago Press.

Herrick C. J. (1948) *The Brain of the Tiger Salamander.* Chicago: University of Chicago Press.

Herrnstein, R. J. (1979) Acquisition, generalization, and discrimination reversal of a natural concept. *Animal Behavior Processes* 5: 116–129.
Herrnstein, R. J., and Loveland, D. H. (1964) Complex visual concept in the pigeon. *Science* 146: 549–551.
Hetzler, B. E., and Griffin, J. L. (1981) Infantile autism and the temporal lobe of the brain. *Journal of Autism and Developmental Disorders* 11: 317–330.
Heyes, C. M. (1993a) Cues, convergence, and a curmudgeon: A reply to Povinelli. *Animal Behaviour* 48: 242–244.
Heyes, C. M. (1993b) Imitation, culture, and cognition. *Animal Behaviour* 46: 999–1010.
Heyes, C. M. (1998) Theory of mind in nonhuman primates. *Behavioral and Brain Sciences* 21: 101–148.
Heyes, C. M., and Dickinson, A. (1990) The intentionality of animal action. *Mind and Language* 5 (1): 87–104.
Heyes, C. M., and German, T. P. (1994) Eye to eye, but not a meeting of minds. *Cahiers de Psychologie Cognitive* 13: 607–614.
Hietanen, J. K., and Perrett, D. I. (1996). Motion sensitive cells in the macaque superior temporal polysensory area: Response discrimination between self-generated and externally generated pattern motion. *Behavioral Brain Research* 76: 155–197.
Hiley, D. R., Bohman, J. F., and Shusterman, R. (1991) *The Interpretative Turn*. Ithaca, N.Y.: Cornell University Press.
Hinde, R. A. (1970) *Animal Behaviour: A Synthesis of Ethology and Comparative Psychology*. 2d ed. London: McGraw-Hill.
Hirschfeld, L. A. (1996) *Race in the Making: Cognition, Culture, and the Child's Construction of Human Kinds*. Cambridge, Mass.: MIT Press.
Hirschfeld, L. A., and Gelman, S. A. ([1994] 1998) *Mapping the Mind*. Reprint, Cambridge: Cambridge University Press.
Hobbes, T. (1651) *Leviathan*. London: Dent and Sons.
Hobson, P. (1984) Early childhood autism and the question of egocentrism. *Journal of Autism and Developmental Disorders* 14: 85–104.
Hobson, P. (1986) The autistic child's appraisal of expressions of emotion. *Journal of Child Psychology and Psychiatry* 27: 321–342.
Hobson, P. (1993) *Autism and the Development of Mind*. Hillsdale, N.J.: Lawrence Erlbaum.
Holt, D. J., Graybiel, A. M., and Saper, C. B. (1997) Neurochemical architecture of the human striatum. *Journal of Comparative Neurology* 384: 1–25.
Horgan, T, and Tienson, J. (1996) *Connectionism and the Philosophy of Psychology*. Cambridge, Mass.: MIT Press.
Houk, J. C., Davis, J. L., and Beiser, D. G. (1995) *Models of Information Processing in the Basal Ganglia*. Cambridge, Mass.: MIT Press.
Hughes, C., and Cutting, A. L. (1999) Nature, nurture, and individual differences in early understanding of mind. *Psychological Science* 5: 429–432.

Hughes, C., Russell, J., and Robbins, T. W. (1994) Evidence for executive dysfunction in autism. *Neuropsychologia* 32: 477–492.

Hume, D. ([1739] 1969) *A Treatise of Human Nature.* Edited by E. C. Mossner. London: Penguin Books.

Humphrey, N. K. (1976) The social function of intellect. In *Growing Points in Ethology,* edited by P. Bateson and R. A. Hinde. Cambridge: Cambridge University Press.

Humphrey, N. K. (1992) *A History of the Mind: Evolution and the Birth of Consciousness.* New York: Simon and Schuster.

Humphreys, G. W., Riddoch, M. J., and Quinlan, P. T. (1988) Casacade processes in picture identification. *Cognitive Neuropsychology* 5: 67–103.

Husserl, E. ([1913] 1931) *Ideas: General Introduction to Pure Phenomenology.* Translated by W. P. Boyce Gibson. London: Geoge Allen and Unwin.

Hutchins, E. (1996) *Cognition in the Wild.* Cambridge, Mass.: MIT Press.

Ishai, A., Ungerleider, L. G., Martin, A., Schouten, J. L., and Haxby, J. V. (1999) Distributed representations of objects in the human visual pathway. *Proceedings of the National Academy of Sciences* 96: 9379–9384.

Insel, T. R. (1992): Oxytocin: A neuropeptide for affiliation—Evidence from behavioral, receptor autoradiographic, and comparative studies. *Psychoneuroendocrinology* 17: 3–33.

Insel, T. R., O'Brien, D. J., and Leckman, J. F. (1999) Oxytocin, vasopressin and autism: Is there a connection? *Biological Psychiatry* 45: 145–147.

Izard, C. E. (1971). *The Face of Emotion.* New York: Appelton-Century-Crofts.

Izard, C. E. (1994) Innate and universal facial expressions: Evidence from developmental and cross-cultural research. *Psychological Bulletin* 115: 288–289.

Jackson, J. H. ([1884] 1958) *Selected Writings of John Hughlings Jackson.* 2 vols. Edited by J. Tayler, G. Holmes, and F. M. R. Walshe. London: Staples.

Jaeger, J. J., Lockwood, A. H., Kemmerer, D. L., Jr., R. D. V. V., Murphy, B. W., and Khalak, H. G. (1996). A positron emission tomographic study of regular and irregular verb morphology in English. *Language* 72: 451–497.

James, W. ([1890] 1952) *The Principles of Psychology.* Reprint New York: Dover.

James, W. ([1907] 1959) *Pragmatism.* Reprint New York: Meridian Books.

James, W. ([1912] 1958) *Essays in Radical Experience.* Reprint New York: Longman, Green.

Jaspers, K. ([1913] 1997) *General Psychopathology.* Reprint, Baltimore: Johns Hopkins University Press.

Jeannerod, M. (1994) The representing brain: Neural correlates of motor intention and imagery. *Behavioral and Brain Sciences* 17: 187–201.

Jeannerod, M. (1997) *The Cognitive Neuroscience of Action.* Oxford: Blackwell.

Jeannerod, M. (1999) To act or not to act: Perspectives on the representations of actions. *Quarterly Journal of Experimental Psychology* 52: 1–29.

Jeannerod, M., and Decety, J. (1995) Mental motor imagery: A window into the representational stages of action. *Current Opinion in Neurobiology* 5: 727–832.

Jenkins, J. M., and Astington, J. W. (1996) Cognitive factors and family structure associated with theory of mind developing in young children. *Developmental Psychology* 32: 70–88.

Jennings, H. J. ([1905] 1962) *Behaviors of the Lower Organisms.* Reprint, Bloomington: Indiana University Press.

Joanisse, M. F., and Seidenberg, M. S. (1999). Impairments in verb morphology after brain injury: A connectionist model. *Proceedings of the National Academy of Sciences USA* 96: 7592–7597.

Joas, H. (1993) *Pragmatism and Social Theory.* Chicago: University of Chicago Press.

Joas, H. (1996) *The Creativity of Action.* Chicago: University of Chicago Press.

Johnson, D. M., and Erneling, C. E. (1997) *The Future of the Cognitive Revolution.* Oxford: Oxford University Press.

Johnson, M. (1987) *The Body on the Mind: The Bodily Basis of Meaning, Imagination, and Reason.* Chicago: University of Chicago Press

Johnson, S. C., and Carey, S. (1998) Knowledge enrichment and conceptual change in folkbiology: Evidence from Williams syndrome. *Cognitive Psychology* 37: 156–200.

Johnston, J. B. (1923) Further contributions to the study of the evolution of the forebrain. *Journal of Comparative Neurology* 5: 337–381.

Jolly, A. (1966) Lemur social behavior and primate intelligence. *Science* 153: 501–506.

Jolly, A. (1985) *The Evolution of Primate Behavior.* New York: Macmillan.

Kagan, J. (1979) *The Growth of the Child: Reflections on Human Development.* Brighton: Harvester.

Kagan, J. (1981) *The Second Year: The Emergence of Self-Awareness.* Cambridge, Mass.: Harvard University Press.

Kagan, J. (1984) *The Nature of the Child.* New York: Basic.

Kahneman, D., and Tversky, A. (1973) On the psychology of prediction. *Psychological Review* 80: 237–251.

Kandel, E. R. (1976) *Cellular Basis of Behavior.* San Francisco: Freeman.

Kanner, L. (1943) Autistic disturbances of affective contact. *Nervous Child* 2: 217–250.

Kanner, L. (1943) Autistic disturbances of affective contact. *Nervous Child* 2: 217–250.

Kanner, L. (1946) Irrelevant and metaphorical language in early infantile autism. *American Journal of Psychology* 103: 242–246.

Kant, I. ([1787] 1929) *Critique of Pure Reason.* Translated by N. K. Smith. New York: St. Martin's Press.

Karakashian, S. J., Gyger, M., and Marler, P. (1988) Audience effects on alarm calling in chickens *(Gallus gallus). Journal of Comparative Psychology* 102: 129–135.

Karmiloff-Smith, A. (1988) The child is a theoretician, not an industivist. *Mind and Language* 3: 183–195.

Karmiloff-Smith, A. (1992) *Beyond Modularity: A Developmental Perspective on Cognitive Science.* Cambridge, Mass.: MIT Press.

Karmiloff-Smith, A., Klima, E., Bellugi, U., Grant, J., and Baron-Cohen, S. (1995) Is there a social module? Language, face processing, and theory of mind in individuals with Williams syndrome. *Journal of Cognitive Neuroscience* 7: 196–208.

Kates, W. R., Mostofsky, S. H., Zimmerman, A. W., Mazzocco, M. M., Landa, R., Warsofsky, I. S., Kaufmann, W. E., and Reiss, A. L. (1998) Neuroanatomical and neurocognitive differences in a pair of monozygous twins discordant for strictly defined autism. *Annals of Neurology* 43: 782–891.

Kaye, K. (1982) *The Mental and Social Life of Babies.* Chicago: University of Chicago Press.

Keil, F. C. (1992) *Concepts, Kinds, and Cognitive Development.* Cambridge, Mass.: MIT Press.

Kelley, A. E. (1999) Neural integrative activities of nucleus accumbens subregions in relation to learning and motivation. *Psychobiology* 27: 198–213.

Kellog, W. N., and Kellog, L. A. (1933) *The Ape and the Child.* New York: McGraw-Hill.

Kelly, K. (1996) *The Logic of Reliable Inquiry.* Oxford: Oxford University Press.

King, B. J. (1986) Extractive foraging and the evolution of primate intelligence. *Human Evolution* 1: 361–372.

King, B. J. (1991) Social information transfer in monkeys, apes, and hominids. *Yearbook of Physical Anthropology* 34: 97–115.

Kinnaman, A. J. (1902) Mental life of two *(Macacus rhesus)* monkeys in captivity. *American Journal of Psychology* 13: 98–148.

Kitcher, P. (1990) *Kant's Transcendental Psychology.* New York: Oxford University Press.

Kitcher, P. (1993) *The Advancement of Science.* New York: Oxford University Press.

Klin, A., Volkmar, F. R., and Sparrow, S. S. (1992) Autistic social dysfunction: Some limitations of the theory of mind hypothesis. *Journal of Child Psychology and Psychiatry* 33: 861–867.

Kling, A. S., and Brothers, L. A. (1992) The amygdala and social behavior. In *The Amygdala,* edited by J. P. Aggleton. New York: Wiley-Liss.

Kluver, H. (1933) *Behavior Mechanisms in Monkeys.* Chicago: University of Chicago Press.

Kluver, H., and Bucy, P. C. (1939) Preliminary analysis of function of the temporal lobes in monkeys. *Archives of Neurology and Psychology* 42: 979–1000.

Knowlton, B. J., Mangels, J. A., and Squire, L. R. (1996) A neostriatal habit learning system in humans. *Science* 273: 1399–1402.

Koch, S. (1993) "Psychology" or the "psychology studies." *American Psychologist* 8: 902–904.

Kohler, W. (1925) *The Mentality of Apes.* London: Routledge and Kegan Paul.

Konorski, J. (1967) *Integrative Activity of the Brain.* Chicago: University of Chicago Press.

Kornblith, H. (1985) *Naturalizing Epistemology.* Cambridge, Mass.: MIT Press.

Kornblith, H. (1993) *Inductive Inference and Its Natural Ground: An Essay in Naturalistic Epistemology.* Cambridge Mass.: MIT Press.

Koslowski, B. (1996) *Theory and Evidence.* Cambridge, Mass.: MIT Press.

Kosslyn, S. M. (1980) *Image and Mind.* Cambridge, Mass.: Harvard University Press.

Kosslyn, S. M. (1994) *Image and Brain: The Resolution of the Imagery Debate.* Cambridge, Mass.: MIT Press.

Kosslyn, S. M., Alpert, N. M., and Thompson, W. L. (1993) Visual mental imagery and visual perception: PET studies. In *Functional MRI of the Brain.* Society for Magnetic Resonance Imaging. Virginia: Arlington.

Kosslyn, S. M., and Anderson, R. A. ([1992] 1995) *Frontiers in Cognitive Neuroscience.* Reprint, Cambridge, Mass.: MIT Press.

Kosslyn, S. M., Pascual-Leone, A., Felician, O., Camposano, S., Keenan, J. P., Thompson, W. L., Ganis, G., Sukel, K. E., and Alpert, N. M. (1999) The role of area 17: I. Visual imagery: Convergent evidence from PET and rTMS. *Science* 284: 167–170.

Kretek J. E., and Price, J. L. (1978) Amygdaloid projections to subcortical structures within the basal forebrain and brainstem in the rat and the cat. *Journal of Comparative Neurology* 178: 225–254.

Kripke, S. A. (1980) *Naming and Necessary.* Cambridge, Mass.: Harvard University Press.

Kripke, S. A. (1982) *Wittgenstein on Rules and Private Language.* Cambridge, Mass.: Harvard University Press.

Kuhn, T. S. ([1962] 1970) *The Structure of Scientific Revolutions.* Reprint, Chicago: University of Chicago Press.

Kummer, H. (1982) Social knowledge in free-ranging primates. In *Animal Mind—Human Mind,* edited by D. R. Griffin. Berlin: Springer.

Kummer, H. (1995) *In Quest of the Sacred Baboon.* Princeton, N.J.: Princeton University Press.

Kunda, Z. (1999) *Social Cognition.* Cambridge, Mass.: MIT Press.

Lakoff, G. (1987) *Fire, Women, and Dangerous Things: What Categories Reveal about the Mind.* Chicago: University of Chicago Press.

Lakoff, G., and Johnson, M. (1999) *Philosophy in the Flesh.* New York: Basic.

Lane, R. D., Nadel, L., Allen, J. J. B., and Kaszniak, A. W. (1999) The study of emotion from the perspective of cognitive neuroscience. In *Cognitive Neuroscience of Emotion,* edited by R. D. Lane and L. Nadel. Oxford: Oxford University Press.

Langer, S. K. (1957) *Philosophy in a New Key.* 3rd ed. Cambridge, Mass.: Harvard University Press.

Lashley, K. S. (1938) An experimental analysis of instinctive behavior. *Psychological Review* 45: 445–471.

Lashley, K. S. (1951) The problem of serial order in behavior. In *Cerebral Mechanisms in Behavior.* New York: Wiley and Sons.
Lashley, K. S. (1958) Cerebral organization and behavior. *Proceedings of the Association for Research in Nervous and Mental Diseases* 36: 1–18.
Leavens, D. A., and Hopkins, W. D. (1998) Intentional communication by chimpanzees: A cross-sectional study of the use of referential gestures. *Developmental Psychology* 34: 813–822.
LeBoyer, M., Bourvard, M. P., Launay, J. M., Tabuteau, F., Waller, D., Dugas, M., Kerdelhue, B., Lensing, P., and Panksepp, J. (1992) A double blind study of naltrexone in infantile autism. *Journal of Autism and Developmental Disorders* 22: 309–319.
Lederhendler, I., and Schulkin, J. (n.d.) *Behavioral Neurosciences: Meeting the Challenge of Genomics.* Unpublished manuscript.
LeDoux, J. E. (1987) Emotion. In *Handbook of Physiology: The Nervous System.* Washington, D.C.: American Physiological Society.
LeDoux, J. E. (1993) Congition versus emotion, again—this time in the brain: A response to Parrott and Schulkin. *Cognition and Emotion* 7: 61–64.
LeDoux, J. E. (1996) *The Emotional Brain.* New York: Simon and Schuster.
Lee, K. Eskritt, M., Symons, L. A., and Muir, D. (1998) Children's use of triadic eye gaze information for "mind reading." *Developmental Psychology* 34: 525–539.
Leekam, S. R., and Perner, J. (1991) Does the autistic child have a metarepresentational deficit? *Cognition* 40: 203–218.
Leibniz, G. W. ([1690] 1982) *New Essays on Human Understanding.* Translated by P. Remnant and J. Bennett. New York: Cambridge University Press.
Leslie, A. M. (1987) Pretense and representation: The origins of "theory of mind." *Psychological Review* 94: 412–426.
Leslie, A. M. (1990) Pretence, autism, and the basis of "theory of mind." *British Psychology Society* 3: 120–123.
Leslie, A. M. (1994) ToMM, ToBY, and agency: Core architecture and domain specifity. In *Mapping the Mind,* edited by L. A. Hirshfield and S. A. Gelman. Cambridge: Cambridge University Press.
Leslie, A. M., and Frith, U. (1989) Autistic children's understanding of seeing, knowing, and believing. *British Journal of Developmental Psychology* 6: 315–324.
Leslie, A. M., and Frith, U. (1990) Prospects for a cognitive neuropsychology of autism: Hobson's choice. *Psychological Review* 97: 122–131.
Leslie, A. M., German, T. P., and Happé, F. G. (1993) Even a theory-theory needs information processing: ToMM, an alternative theory-theory of the child's theory of mind. *Behavioral and Brain Sciences* 16: 1.
Leslie, A. M., and Happe, F. (1989) Autism and ostensive communication: The relevance of metarepresentation. *Development and Psychopathology* 1: 205–212.

Leslie, A. M., and Roth, D. ([1993] 1999) What autism teaches us about metarepresentation. In *Understanding Other Minds,* edited by S. Baron-Cohen, H. Tager-Flusberg, and D. J. Cohen. Reprint, Oxford: Oxford University Press.

Leslie, A. M., and Thaiss, L. (1992) Domain specificity in conceptual development: Neuropsychological evidence from autism. *Cognition* 43: 225–251.

Levenson, R. W., Ekman, P., Heider, K., and Friesen, W. V. (1992) Emotion and autonomic nervous system activity: I. The Minangkabau of West Sumatra. *Journal of Personality and Social Psychology* 62: 972–988.

Lewis, C. I. ([1929] 1956) *Mind and the World Order: Outline of a Theory of Knowledge.* Reprint, New York: Dover.

Lewis, D. (1972) Psychophysical and theoretical identification. *Australasian Journal of Philosophy* 50: 249–258. Reprinted in N. Block, ed. (1980) *Readings in Philosophy of Psychology,* vol. 1. London: Methuen.

Lewis, D. (1983) *Philosophical Papers,* vol. 1. Oxford: Oxford University Press.

Lieberman, M. D. (2000) Intuition: A social cognitive neuroscience approach. *Psychological Bulletin,* in press.

Lillard, A. S. (1997) Other folks' theories of mind and behavior. *Psychological Science* 8: 268–274.

Lillard, A. (1998) Ethnopsychologies: Reply to Wellman and Gauvain. *Psychological Bulletin* 23: 43–46.

Lillard, A. S. (1999) Developing a cultural theory of mind. *Psychological Science* 28: 57–62.

Livingston, K. R., Andrews, J. K., and Harnard, S. (1998) Categorical perception effects induced by category learning. *Journal of Experimental Psychology* 24: 732–853.

Lloyd-Morgan, C. (1930) *The Animal Mind.* London: Edward Arnold.

Loar, B. (1981) *Mind and Meaning.* Cambridge: Cambridge University Press.

Locke, J. ([1690] 1964) *An Essay Concerning Human Understanding.* Edited by J. Yolton. London: Dutton.

Loeb, J. ([1918] 1973). *Forced Movements, Tropisms, and Animal Conduct.* Reprint, New York: Dover.

Lorenz, K. (1965) *Evolution and the Modification of Behavior.* Chicago: University of Chicago Press.

Loveland, K. A. (1993) Autism, affordances, and the self. In *The Perceived Self,* edited by U. Neisser. Cambridge: Cambridge University Press.

Luria, A. R. (1973) *The Working Brain.* New York: Basic.

Lycan, W. G. (1996) *Consciousness and Experience.* Cambridge, Mass.: MIT Press.

Lyons, W. (1995) *Approaches to Intentionality.* Oxford: Clarendon.

Mackintosh, N. J. (1975) A theory of attention: Variations in the associability of stimuli with reinforcement. *Psychological Review* 82: 276–298.

MacLean, P. D. (1990) *The Triune Brain in Evolution.* New York: Plenum.

Maier, N. R. F., and Schneirla, T. C. (1925) *Principles of Animal Psychology.* New York: McGraw-Hill.

Markman, E. M. (1989) *Categorization and Naming in Children: Problems of Induction*. Cambridge, Mass.: MIT Press.

Marler, P. (1961) The logical analysis of animal communication. *Journal of Theoretical Biology* 1: 295–317.

Marler, P. (1976) Social organization, communication, and graded signals: The chimpanzee and the gorilla. In *Growing Points in Ethology*, edited by P. P. G. Bateson and R. A. Hinde. Cambridge: Cambridge University Press.

Marler, P. (1990a) Innate learning preferences: Signals for communication. *Developmental Psychobiology* 23: 557–568.

Marler, P. (1990b) Song learning: The interface between behaviour and neuroethology. *Philosophical Transactions of the Royal Society of London—Series B: Biological Sciences* 329: 109–114.

Marler, P. (n.d.) Review of apes, language, and the human mind. *American Anthropologist*. In press.

Marler, P., Dufty, A., and Pickert, R. (1986a) Vocal communication in the domestic chicken: 1. Does a sender communicate information about the quality of a food referent to a receiver? *Animal Behaviour* 34: 188–193.

Marler, P., Dufty, A., and Pickert, R. (1986b) Vocal communication in the domestic chicken: 2. Is a sender sensitive to the presence and nature of a receiver? *Animal Behaviour* 34: 194–198.

Marler, P., Evans, C., and Hauser, M. (1992) Animal signals: Motivational, referential, or both? In *Nonverbal Communication: Comparative and Developmental Approaches*, edited by H. Papousek, U. Jurgens, and M. Papousek. Cambridge: Cambridge University Press.

Marler, P., and Hamilton, W. J., III (1966) *Mechanisms of Animal Behavior*. New York: John Wiley and Sons.

Marler, P., and Peters, S. (1982) Developmental overproduction and selective attrition: New processes in the epigenesis of birdsong. *Developmental Psychobiology* 15 (4): 369–378.

Marr, D. (1982) *Vision*. San Francisco: Freedman.

Marsden, C. D. (1984a) Function of the basal ganglia as revealed by cognitive and motor disorders in Parkinson's disease. *Canadian Journal of Neurology* 11: 129–135.

Marsden, C. D. (1984b) The pathophysiology of movement disorders. *Neurological Clinics* 2: 435–459.

Martin, A. (1987) Representation of semantic and spatial knowledge in Alzheimer's patients: Implications for models of preserved learning in amnesia. *Journal of Clinical and Experimental Neuropsychology* 9: 191–224.

Martin, A. (1998a) Automatic activation of the medial temporal lobe during encoding: Lateralized influences of meaning and novelty. *Hippocampus* 9: 61–80.

Martin, A. (1999b) Organization and origins of semantic knowledge in the brain. In *The Origin and Diversification of Language*, edited by N. G. Jablonski. San Francisco: California Academy of Sciences.

Martin, A., Haxby, J. V., LaLonde, F. M., Wiggs, C. L., and Ungerleider, L. G. (1995) Discrete cortical regions associated with knowledge of color and knowledge of action. *Science* 270: 102–105.

Martin, A., Wiggs, C. L., Ungerleider, L. G., and Haxby, J. V. (1996) Neural correlates of category-specific knowledge. *Nature* 379: 649–651.

Mathews, G. B. (1980) *Philosophy and the Young Child*. Cambridge, Mass.: Harvard University Press.

Mayr, E. (1960). The emergence of evolutionary novelties. In *Evolution after Darwin*, vol. 1, edited by S. Tax. Chicago: University of Chicago Press.

McCarthy, R. A., and Warrington, E. K. (1988) Evidence for modality-specific meaning systems in the brain (see comments). *Nature* 334: 428–430.

McCarthy, R. A., and Warrington, E. K. (1994) Disorders of semantic memory. *Philosophical Transactions of the Royal Society of London—Series B: Biological Sciences* 346 (1315): 89–96.

McClelland, J. L., and Rumelhart, D. E. (1986) *Parallel Distributed Processing*. Cambridge, Mass.: MIT Press.

McEwen, B. S., and Schmeck, H. M. (1994) *The Hostage Brain*. New York: Rockefeller University Press.

McFarland, D. (1991) Defining motivation and cognition in animals and robots. *International Studies in the Philosophy of Science* 5: 153–170.

McGinn, C. (1989) *Mental Content*. Oxford: Basil Blackwell.

McGinn, C. (1991) *The Problem of Consciousness: Essays Towards a Resolution*. Oxford: Basil Blackwell.

McGinty, J. F., ed. (1999) *Advancing from the ventral striatum to the extended amygdala*. New York: New York Academy of Sciences.

Mead, G. H. (1932) *The Philosophy of the Present*. Edited by Arthur E. Murphy. Chicago: University of Chicago Press.

Mead, G. H. (1934) *Mind, Self, and Society*. Edited by Charles W. Morris. Chicago: University of Chicago Press.

Mead, G. H. (1938) *The Philosophy of the Act*, vol. 3. Edited by C. W. Morris. Chicago: University of Chicago Press.

Meltzoff, A. N. (1995) Understanding the intentions of others: Re-enactment of intended acts by 18-month-old children. *Developmental Psychology* 31: 838–850.

Meltzoff, A. N., and Moore, M. (1994) Imitation, memory, and the representation of persons. *Infant Behavior and Development* 17: 83–99.

Menzel, E. W., Jr., Davenport, R. K., and Rogers, C. M. (1972) Protocultural aspects of chimpanzees' responsiveness to novel objects. *Folia Primatologica* 17: 161–170.

Menzel, E. W., Jr., and Juno, C. (1985) Social foraging in marmoset monkeys and the question of intelligence. *Philosophical Transactions of the Royal Society of London—Series B* 308: 145–158.

Merleau-Ponty, M. ([1942] 1967) *The Structure of Behavior.* Translated by A. L. Fisher. Boston: Beacon Press.

Merleau-Ponty, M. ([1962] 1994) *Phenomenology.* Translated by C. Smith. New York: Routledge.

Merleau-Ponty, M. (1964) *The Primacy of Perception.* Edited and with an introduction by J. M. Edie. Chicago: Northwestern University Press.

Miller, G. A., Galanter, E., and Pribram, K. H. (1960) *Plans and the Structure of Behavior.* New York: Holt, Rinehart and Winston.

Miller, N. E. (1959) Chemical coding in the brain. *Science* 148: 328–338.

Miller, N. E. (1971) *Selected Papers,* vols. 1 and 2. Chicago: Aldine and Atherton.

Millikan, R. G. (1984) *Language, Thought, and Other Biological Categories: New Foundations for Realism.* Cambridge, Mass.: MIT Press.

Millikan R. G. (1998) A common structure for concepts of individuals, stuffs, and real kinds: More mama, more milk, and more mouse. *Behavioral and Brain Sciences* 21: 55–100.

Milner, B., and Petrides, M. (1984) Behavioral effects of frontal-lobe lesion in man. *Trends in Neuroscience* 7: 403–407.

Milner, P. M., and White, N. M. (1987) What is physiological psychology? *Psychobiology* 15: 2–6.

Mineka, S., and Cook, M. (1988) Social learning and the acquisition of snake fear in monkeys. In *Social Learning: Psychological and Biological Perspectives,* edited by T. R. Zebntall and B. G. Galef Jr. Hillsdale, N.J.: Lawrence Erlbaum.

Mineka, S., and Cook, M. (1993) Mechanisms involved in the observational conditioning of fear. *Journal of Experimental Psychology: General* 122: 23–28.

Mishkin, M., Malamut, B., and Bachevalier, J. (1984) Memories and habits: Two neural systems. In *Neurobiology of Learning and Memory,* edited by J. L. McGaugh, G. Lynch, and N. M. Weinberger. New York: Gulliford.

Mishkin, M., Suzuki, W. A., Gadian, D. G., and Vargha-Khadem, F. (1997) Hierarchical organization of cognitive memory. *Philosophical Transactions of the Royal Society of London* 352: 1461–1467.

Mithen, S. (1990). *Thoughtful Foragers: A Study of Prehistoric Decision Making.* Cambridge: Cambridge University Press.

Mithen, S. (1996). *The Prehistory of the Mind: The Cognitive Origins of Art, Religion, and Science.* London: Thames and Hudson.

Modahl, C., Green, L. A., Fein, D., Morris, M., Waterhouse, L., Feinstein, C., and Levin, H. (1998) Plasma oxytocin levels in autistic children. *Biological Psychiatry.* 43: 270–278.

Moore, C. (1996) Theories of mind in infancy. *British Journal of Developmental Psychology* 14: 19–40.

Moore, C., and Corkum, V. (1994) Social understanding at the end of the first year of life. *Developmental Review* 14: 349–382.

Moore, C., and Dunham, P. J., eds. (1995) *Joint Attention.* Hillsdale, N.J.: Lawrence Earlbaum.

Morgan, C., and Stellar, E. (1950) *Physiological Psychology.* 2d ed. New York: McGraw-Hill.

Morris, C. W. (1938) *Foundations of the Theory of Signs.* Chicago: University of Chicago Press.

Morris, J. S., Fritgh, C. D., Perrett, D. I., Rowland, D., Young, A. W., Cadler, A. J., and Dolan, R. J. (1996) A differential neural response in the human amygdala to fearful and happy facial expressions. *Nature* 383: 812–815.

Moscovitch, M., Behmann, M., and Winocur, G. (1994) Do PETS have long or short ears? Mental imagery and neuroimaging. *Trends in Neuroscience* 17.

Moses, L. J., and Chandler, M. J. (1992) Traveller's guide to children's theories of mind. *Psychological Inquiry* 3: 286–301.

Mounce, H. O. (1997) *The Two Pragmatisms: From Peirce to Rorty.* New York: Routledge.

Mountcastle, V. B. (1978) An organizing principle for cerebral function: The unit module and the distributed system. In *The Mindful Brain.* Cambridge, Mass.: MIT Press.

Mountz, J. M., Tolbert, L. C., Lill, D. W., Katholi, C. R., and Liu, H. G. (1995) Functional deficits in autistic disorder: Characterization by technetium—99m-HMPAO and SPECT. *Journal of Nuclear Medicine* 36: 1156–1162.

Mundy, P. (1995) Joint attention and social-emotional approach behavior in children with autism. *Development and Psychopathology* 7: 63–82.

Mundy, P., Sigman, M., and Kasart, C. (1994) Joint attention, developmental level, and symptom presentation in autism. *Development and Psychopathology* 5: 389–401.

Nagel, T. (1974) What is it like to be a bat? *Philosophical Review* 83: 435–450.

Nagel, T. (1979) *Mortal Questions.* New York: Cambridge University Press.

Nagel, T. (1986) *The View from Nowhere.* Oxford University Press.

Nagell, K., Olguin, R. S., and Tomasello, M. (1993) Processes of social learning in the tool use of chimpanzees *(Pan troglodytes)* and human children *(Homo sapiens). Journal of Comparative Psychology* 107: 174–186.

Nauta, W. J. H., and Domesick, V. B. (1978) Crossroads of limbic and striatal circuitry: Hypothalamonigral connections. In *Limbic Mechanisms,* edited by K. E. Livingston and O. Hornykiewicz. New York: Plenum.

Neisser, U. (1967) *Cognitive Psychology.* New York: Appleton-Century-Crofts.

Neisser, U. (1984) Toward an ecologically oriented cognitive science. In *New Directions in Cognitive Science,* edited by T. Schlecter and M. Toglia. Norwood, N.J.: Ablex.

Nelson, D. A., and Marler, P. (1989) Categorical perception of a natural stimulus continuum: Birdsong. *Science* 244: 976–978.

Neville, R. C. (1974) *The Cosmology of Freedom.* New Haven, Conn.: Yale University Press.

Neville, R. C. (1989) *Recovery of the Measure: Interpretation and Nature.* New York: State University of New York Press.

Newell, A. (1990) *Unified Theories of Cognition.* Cambridge, Mass.: Harvard University Press.

Nishimori, K., Young, L., Guo, Q., Wang, Z., Insel, T., and Matzuk, M. (1996) Oxytocin is required for nursing but is not essential for parturition or reproductive behavior. *Proceedings of the National Academy of Sciences USA* 93: 777–783.

Norgren, R. (1976) Taste pathways to hypothalamus and amygdala. *Journal of Comparative Neurology* 166: 17–30.

Norgren, R. (1995) Gustatory system. In *The Rat Nervous System,* edited by G. Paxinos. Orlando: Academic.

O'Connell, S. O. (1997) *Mindreading.* New York: Bantam.

Ogden, D. L., Thompson, R. K. R., and Premack, D. (1988) Spontaneous transfer of matching by infant chimpanzees *(Pan troglodytes). Animal Behavior Processes* 14: 140–145.

Ogden, D. L., Thompson, R. K. R., Premack, D. (1990) Infant chimpanzees spontaneously perceive both concrete and abstract same/different relations. *Child Development* 61: 621–631.

Olds, J., and Milner, P. (1954) Positive reinforcement produced by electrical stimulation of septal areas and other regions of rat brains. *Journal of Comparative and Physiological Psychology* 47: 419–427.

O'Neill, D. K., and Povinelli, D. J. (n.d.) Chimpanzees, children, and the coordination of first and third person intentional relations. In review.

Oram, M. W., and Perrett, D. I. (1996). Integration of form and motion in the anterior superior temporal polysensory area (STPa) of the macaque monkey. *Journal of Neurophysiology* 76: 109–129.

Panksepp, J. (1979) A neurochemical theory of autism. *Trends in Neuroscience* 2: 174–177.

Panksepp, J. (1998) *Affective Neuroscience.* Oxford: Oxford University Press.

Panksepp, J. and Lensing, P. (1991) Naltrexone treatment and autism: A synopsis of an open trial with four children. *Journal of Autism and Developmental Disorders* 21: 135–141.

Papez, J. W. (1929) *Comparative Neurology.* New York: Hafner.

Papineau, D. (1993) *Philosophical Naturalism.* Cambridge: Blackwell.

Parker, S. T., and Gibson, K. R. (1979) A developmental model for the evolution of language and intelligence in early hominids. *Behavioral and Brain Sciences* 2: 367–408.

Parrott, W. G., and Sabini, J. (1989) On the emotional quality of certain types of cognition: A reply to arguments for the independence of cognition and affect. *Cognition Therapy and Research* 13: 49–65.

Parrott, W. G., and Schulkin, J. (1993a) Neuropsychology and the cognitive nature of the emotions. *Cognition and Emotion* 7: 43–59.

Parrott, W. G., and Schulkin, J. (1993b) What sort of system could an affective system be? A reply to LeDoux. *Cognition and Emotion* 7: 65–69.

Parsons, L. M., Fox, P. T., Downs, J. H., et al. (1995) Use of implicit motor imagery for visual shape discrimination as revealed by PET. *Nature* 375: 54–58.

Pasculavaca, D. M., Fantie, B. D., Papageorgiou, M., and Mirsky, A. F. (1998) Attentional capacities in children with autism: Is there a general deficit in shifting focus? *Journal of Autism and Developmental Disorders* 28: 467–478.

Pashler, H. E. (1998) *The Psychology of Attention*. Cambridge, Mass.: MIT Press.

Passingham, R. E. (1981) Primate specialization in brain and intelligence. *Symposium of the Zoological Society of London* 46: 361–388.

Passingham, R. E. (1982) *The Human Primate.* San Francisco: W. H. Freeman.

Patterson, F. (1990) Mirror behavior and self-concept in the lowland gorilla. Abstract. *American Journal of Primatology* 20: 219–220.

Pavlov, I. P. (1928) *Lectures on Conditioned Reflexes*. New York: Liverright.

Peirce, C. S. (1871) Fraser's *The Works of George Berkeley* (review). *North American Review* (October): 487–490.

Peirce, C. S. (1877) The fixation of belief. *Popular Scientific Monthly* 12: 1–15.

Peirce, C. S. (1878a) Deduction, induction, and hypothesis. *Popular Science Monthly* (August): 323–338.

Peirce, C. S. (1878b) How to make our ideas clear. *Popular Science Monthly* (January): 286–302.

Peirce, C. S. (1887) Logical machines. *American Journal of Psychology* 1: 165–170.

Peirce, C. S. ([1889] 1992) *Reasoning and the Logic of Things*. Reprint, Cambridge, Mass.: Harvard University Press.

Peirce, C. S. (1992) *The Essential Peirce,* edited by N. Jouser and C. Kloesel. Bloomington: Indiana University Press.

Pepperberg, I. (1990). Conceptual abilities of some non-primate species, with an emphasis on an African Grey parrot. In *Language and Intelligence in Monkeys and Apes,* edited by S. T. Parker and K. R. Gibson. Cambridge: Cambridge University Press.

Perani, D., Cappa, S. F., Bettinardi, V., Bressi, S., Gorno-Tempini, M., Matarrese, M., and Fazio, F. (1995) Different neural systems for the recognition of animals and man-made tools. *Neuroreport* 6: 1637–1641.

Perani, D., Schnur, T., Tettamanti, M., Gorno-Tempini, M., Cappa, S. F., and Fazio, F. (1999) Word and picture matching: A PET study of semantic category effects. *Neuropsychologia* 37: 293–306.

Perner, J. (1991) *Understanding the Representational Mind*. Cambridge, Mass.: MIT Press.

Perner, J. ([1993] 1999) The theory of mind deficit in autism: Rethinking the metarepresentation theory. In *Understanding Other Minds: Perspectives from Autism,* edited by S. Baron-Cohen, H. Tager-Flusberg, and D. J. Cohen. Reprint, New York: Oxford University Press.

Perner, J. (1995) The many faces of belief: Reflections on Fodor's and the child's theory of mind. *Cognition* 57: 241–269.

Perner, J., Leekam, S. R., and Wimmer, H. (1987) Three year olds' difficulty with false belief. *British Journal of Developmental Psychology* 5: 125–137.

Perner, J., Ruffman, T., and Leekam, S. R. (1994) Theory of mind is contagious: You catch it from your sibs. *Child Development* 65: 1228–1238.

Perrett, D. I., and Emery, N. J. (1994). Understanding the intentions of others from visual signals: Neurophysiological evidence. *Cahiers de Psychologie Cognitive* 13: 683–694.

Perrett, D., and Mistlin, A. (1990) Perception of facial characteristics by monkeys. In *Comparative Perception,* volume 2: *Complex Signals,* edited by W. Stebbins and M. Berkeley. New York: Wiley.

Perrett, D. I., Harries, M. H., Mistlin, A. J., and Hietanen, J. K. (1990) Social signals analyzed at the single cell level: Someone is looking at me, something touched me, something moved. *International Journal of Comparative Psychology* 4: 25–55.

Perrett, D. I., Hietanen, J. K., Oram, M. W., and Benson, P. J. (1992) Organization and functions of cells responsive to faces in the temporal cortex. *Philosophical Transactions of the Royal Society of London, B* 335: 23–30.

Peterson, C. C., and Siegal, M. (1999) Representing inner worlds: Theory of mind in autistic, deaf, and normal hearing children. *Psychological Science* 110: 126–130.

Pfaff, D. W. (1999) *Drive.* Cambridge, Mass.: MIT Press.

Philips, W., Baron-Cohen, S., and Rutter, M. (1992) The role of eye-contact in goal-detection: Evidence from normal toddlers and children with autism or mental handicap. *Development and Psychopathology* 4: 374–384.

Piaget, J. ([1929] 1954) *The Construction of Reality in the Child.* Reprint, New York: W. W. Norton.

Piaget, J. (1952) *The Origins of Intelligence in Children.* New York: W. W. Norton.

Piaget, J. (1962) *Play Dreams and Imitation.* New York: W. W. Norton.

Piaget, J. (1971) *Biology and Knowledge: An Essay on the Relations between Organic Regulations and Cognitive Processes.* Chicago: University of Chicago Press.

Piaget, J. (1974) *Understanding Causality.* New York: W. W. Norton.

Picard, R. (1997) *Affective Computing.* Cambridge, Mass.: MIT Press.

Pillow, B. H., and Henrichon, A. J. (1996) There's more to the picture than meets the eye: Young children's difficulty understanding biased interpretation. *Child Development* 67 (3): 803–819.

Pinker, S. (1994) *The Language Instinct.* New York: William Morrow.

Pinker, S. (1997a) *How the Mind Works.* New York: W. W. Norton.

Pinker, S. (1997b) Language as a psychological adaptation. *Ciba Foundation Symposium* 208: 162–182.

Pinker, S., and Prince, A. (1991) Regular and irregular morphology and the psychological status of rules of grammar. *Berkeley Linguistic Society* 17: 230–251.

Piven, J., Berthier, M. L., Starkstein, S. E., Nehme, E., et al. (1990) Magnetic resonance imaging evidence for a defect of cerebral cortical development in autism. *American Journal of Psychiatry* 147: 734–839.

Plato. (1961) *The Collected Dialogues of Plato.* Edited by E. Hamilton and H. Ciarns. Princeton: Princeton University Press.

Plotkin, H. (1993) *Darwin Machines and the Nature of Knowledge.* Cambridge, Mass.: Harvard University Press.

Popper, K. R. ([1959] 1968) *The Logic of Scientific Discovery.* Reprint New York: Harper Torchbooks.

Popper, K. R. (1963) *Conjectures and Refutations: The Growth of Scientific Knowledge.* New York: Harper Torchbooks.

Popper, K. R.. (1972) *Objective Knowledge.* Oxford: Oxford University Press.

Posner, M. I., ed. (1989) *Foundations of Cognitive Science.* Cambridge, Mass.: MIT Press.

Posner, M. I., Inhoff, A. W., Friedrich, F. J., and Cohen, A. (1987) Isolating attention systems: A cognitive-anatomical analysis. *Psychobiology* 15: 107–121.

Poulin-Dubois, D., and Shultz, T. R. (1988) The development of the understanding of human behavior: From agency to intentionality. In *Developing theories of mind,* edited by J. W. Astington, P. L. Harris, and D. R. Olson. New York: Cambridge University Press.

Povinelli, D. J., and Cant, J. G. H. (1995) Arboreal clambering and the evolution of self-conception. *Quarterly Review of Biology* 70: 393–421.

Povinelli, D. J., and Davis, D. R. (1994) Differences between chimpanzees *(Pan troglodytes)* and humans *(Homo sapiens)* in the resting state of the index finger: Implications for pointing. *Journal of Comparative Psychology* 108: 134–139.

Povinelli, D. J., and Eddy, T. J. (1996a) *What Young Chimpanzees Know About Seeing.* Monographs of the Society for Research in Child Development, serial no. 247, 61 (3): 1–152.

Povinelli, D. J., and Eddy, T. J. (1996b) Chimpanzees: Joint visual attention. *American Psychological Society* 7: 129–135.

Povinelli, D. J., and Eddy, T. J. (1997). Specificity of gaze-following in young chimpanzees. *British Journal of Developmental Psychology* 15: 213–222.

Povinelli, D. J., Gallup, G. G., Jr., Eddy, T. J., Bierschwale, D. T., Engstrom, M. C., Perilloux, H. K., and Toxopeus, I. B. (1997) Chimpanzees recognize themselves in mirrors. *Animal Behavior* 53: 1083–1088.

Povinelli, D. J., Perilloux, H. K., Reaux, J. E., and Bierschwale, D. (1998) Young and juvenile chimpanzees' *(Pan troglodytes)* reactions to intentional versus accidental and inadvertent actions. *Behavioral Processes* 42: 205–218.

Povinelli, D. J., and Preuss, T. M. (1995). Theory of mind: Evolutionary history of a cognitive specialization. *Trends in Neurosciences* 18: 418–424.

Povinelli, D. J., Rulf, A. B., Landau, K. R., and Bierschwale, D. T. (1993) Self-recognition in chimpanzees *(Pan troglodytes):* Distribution, ontogeny, and patterns of emergence. *Journal of Comparative Psychology* 107: 347–372.

Premack, D. (1983) The codes of man and beasts. *Behavioral and Brain Sciences* 6: 125–167.

Premack, D. (1990) The infant's theory of self-propelled objects. *Cognition* 36 (1): 1–16.

Premack, D. (1994) Levels of causal understanding in chimpanzees and children. *Cognition* 50 (1–3): 347–362.

Premack, D. (1995) Cause/induced motion: Intention/spontaneous motion. In *Origins of the Human Brain,* edited by J.-P. Changeux and J. Chavaillon. Oxford: Clarendon.

Premack, D., and Dasser, V. (1991) Perceptual origins and conceptual evidence for theory of mind in apes and children. In *Natural Theories of Mind,* edited by A. Whiten. Oxford: Blackwell.

Premack, D., and Premack, A. J. (1983) *The Mind of an Ape.* New York: W. W. Norton.

Premack, D., and Premack, A. J. (1994) Origins of human social competence. In *The Cognitive Neurosciences,* edited by M. S. Gazzaniga. Cambridge, Mass.: MIT Press.

Premack, D., and Woodruff, G. (1978) Does the chimpanzee have a theory of mind? *Behavioral and Brain Sciences* 4: 515–526.

Preuss, T. M. (1994) The argument from animals to humans in cognitive neuroscience. In *The Cognitive Neurosciences,* edited by M. S. Gazzaniga. Cambridge, Mass.: MIT Press.

Prior, M. R., and Hoffman, W. (1990) Neuropsychological testing of autistic children through an exploration with frontal lobe tests. *Journal of Autism and Developmental Disorder* 20: 581–590.

Putnam, H. (1975) The meaning of meaning. In *Language, Mind, and Knowledge,* edited by K. Gunderson. Minnesota Studies in the Philosophy of Science, no. 7. Minneapolis: University of Minnesota Press.

Putnam, H. (1989) *Representation and Reality.* Cambridge, Mass.: MIT Press.

Putnam, H. (1990) *Realism with a Human Face.* Edited by J. Conant. Cambridge, Mass.: Harvard University Press.

Putnam H. (1997) Functionalism: Cognitive science or science fiction? In *The Future of the Cognitive Revolution,* edited by D. M. Johnson and C. E. Erneling). Oxford: Oxford University Press.

Pylyshyn, Z. (1999) Is vision continuous with cognition? The case for cognitive impenetrability of visual perception. *Behavioral and Brain Sciences* 22: 341–423.

Quiatt, D., and Reynolds, V. ([1993] 1995) *Primate Behavior.* Reprint, Cambridge: Cambridge University Press.

Quine, W. V. O. ([1959] 1960) *Word and Object.* Reprint, Cambridge, Mass.: MIT Press.

Quine, W. V. O. (1961) *From a Logical Point of View.* 2d ed. Cambridge, Mass.: Harvard University Press.

Quine, W. V. O. ([1966] 1976) *The Ways of Paradox and Other Essays.* Reprint, Cambridge, Mass.: Harvard University Press.

Quine, W. V. O. (1969) *Ontological Relativity and Other Essays.* New York: Columbia University Press.

Rainbow, P., and Sullivan, W. N. (1979) *Interpretative Social Science.* Berkeley and Los Angeles: University of California Press.

Ratner, C. (1989) A social constructionist critique of the naturalistic theory of emotion. *Journal of Mind and Behavior* 10: 211–230.

Reber, P. G., Knowlton, B. J., and Squire, L. R. (1996) Dissociable properties of memory systems: Differences in flexibility of declarative and nondeclarative knowledge. *Behavioral Neuroscience* 110: 861–871.

Regier, T. (1996) *The Human Semantic Potential Spatial Language and Constrained Connectionism.* Cambridge, Mass.: MIT Press.

Reid, T. (1969) *Essays on the Intellectual Powers of Man.* Cambridge, Mass.: MIT Press.

Rescorla, R. A. (1967) Pavlovian conditioning and its proper control procedures. *Psychological Review* 74: 71–80.

Rescorla, R. A., and Wagner, A. R. (1972) A theory of Pavlovian conditioning: Variations in the effectiveness of reinforcement and nonreinforcement. In *Classical Conditioning,* edited by A. H. Black and W. F. Prokasy. New York: Appleton-Century-Crofts.

Rey, G. (1997) *Contemporary Philosophy of Mind: A Contentiously Classical Approach.* Cambridge, Mass.: Blackwell.

Rey, G. (1999) Physicalism and psychology: A plea for substantive philosophy of mind. In *Physicalism,* edited by B. Loewer. Cambridge: Cambridge University Press.

Richter, C. P. (1942) Physiological psychology. *Annual Review of Physiology* 4: 561–574.

Ring, H. A., Baron-Cohen, S., Wheelwright, S., Williams, S. C. R., Brammer, M., Andrew, C., and Bullmore, E. T. (1999) Cerebral correlates of preserved cognitive skills in autism: A functional Magnetic Resonance Imaging (MRI)) study of embedded figures task performance. *Brain* 122: 1305–1315.

Ristau, C. A. (1990) Aspects of the cognitive ethology of an injury feigning plover. In *Cognitive Ethology: The Minds of Other Animals,* edited by C. A. Ristau. Hillsdale, N.J.: Lawrence Erlbaum.

Ristau, C. A. (1998) Cognitive ethology: The minds of children and animals. In *The Evolution of Mind,* edited by D. D. Cummins and C. Allen. Oxford: Oxford University Press.

Rizzolatti, G., and Arbib, M. A. (1998) Language within our grasp. *Trends in Neuroscience* 21: 188–194.

Robinson, D. N. (1993) Is there a Jamesian tradition in psychology? *American Psychologist* 48: 638–643.

Robinson, J. K., and Woodward, W. R. (1989) The convergence of behavioral biology and operant psychology: Toward an interlevel and interfield science. *Behavioral Analyst* 12: 131–141.

Rogers, S. J., Benetto, L., McEvoy, R., and Pennington, B. F. (1996) Imitation and pantomime in high-functioning adolescents with autism spectrum disorders. *Child Development* 67: 2060–2083.

Roland, P. E., and Gulyas, B. (1994) Visual imagery and visual representation. *Trends in Neurosciences* 17: 281–287.

Rolls, E. T. (1996) The orbitofrontal cortex. *Philosophical Transactions of the Royal Society* 351: 1433–1443.

Rolls, E. T. (1999) *The Brain and Emotion*. Oxford: Oxford University Press.

Rolls, E. T., and Treves, A. (1998) *Neural Networks and Brain Function*. Oxford: Oxford University Press.

Rolls, E. T., Treves, A., and Tovee, M. T. (1997) The representational capacity of the distributed encoding of information provided by populations of neurons in primate visual cortex. *Experimental Brain Research* 114: 149–162.

Romanes, G. J. (1882) *Mental Evolution in Animals*. London: Kegan, Paul Trench.

Rorty, R. (1979) *Philosophy and the Mirror of Nature*. Princeton: Princeton University Press.

Rorty, R. (1982) *Consequences of Pragmatism (Essays: 1972–1980)*. Minneapolis: University of Minnesota Press.

Rorty, R. (1991) *Objectivity, Relativism, and Truth: Philosophical Papers*, vol. 1. New York: Cambridge University Press.

Rosch, E. (1973) On the internal structure of perceptual and semantic categories. In *Cognitive Development and the Acquisition of Language*, edited by T. Moore. New York: Academic.

Rosch, E., and Heider, E. (1973) Natural categories. *Cognitive Psychology* 4: 328–350.

Rosch, E., Mervis, C. B., Gray, W. D., Johnson, D. M., and Boyes-Braem, P. (1976) Basic objects in natural categories. *Cognitive Psychology* 8: 382–439.

Rosen, J. B., and Schulkin, J. (1998) From normal fear to pathological anxiety. *Psychological Review* 105: 325–350.

Rosenfeld, S. A., and Van Hoesen, G. W. (1979) Face recognition in the rhesus monkey. *Neuropsychologia* 17: 503–509.

Roth, D., and Leslie, A. M. (1991) The recognition of attitude conveyed by utterance: A study of preschool and autistic children. *British Journal of Development Psychology* 9: 315–330.

Rozin, P. (1976) The evolution of intelligence and access to the cognitive unconscious. In eds. *Progress in Psychobiology and Physiological Psychology*, edited by J. M. Sprague and A. N. Epstein. New York: Academic.

Rozin, P. (1998) Evolution and development of brain and cultures: Some basic principles and interactions. In *Brain and Mind: Evolutionary Perspectives*, edited by M. S. Gazzaniga and J. S. Altman. Strasbourg, France: Human Frontiers Science Program.

Rozin, P., and Kalat, J. (1971) Specific hungers and poison avoidance as adaptive specializations of learning. *Psychological Review* 78: 459–486.

Rumbaugh, D. M. (1995) Primate language and cognition: Common ground. *Social Research* 62: 711–830.

Rumelhart, D. E. (1989) The architecture of mind: A connectionist approach. In *Foundations of Cognitive Science*. Cambridge, Mass.: MIT Press.

Rumsey, J. M., and Hamburger, S. D. (1988) Neuropsychological findings in high-functioning men with infantile autism, residual state. *Journal of Clinical Experimental Neuropsychology* 10: 201–221.

Runciman, W. G., Maynard-Smith, J., and Dunbar, R. I. M. ([1996] 1998) *Evolution of Social Behavior Patterns in Primates and Man*. Reprint, Oxford: Oxford University Press.

Russell, B. ([1900] 1949) *A Critical Exposition of the Philosophy of Leibniz*. Reprint, London: George Allen and Unwin.

Russell, B. ([1950] 1980) An inquiry into meaning and truth: The William James lectures for 1940 delivered at Harvard University. Reprint, Boston: UNWIN Paperbacks.

Russell, C. I. (1994) Is there universal recognition of emotion from facial expression: A review of cross-cultural studies. *Psychological Bulletin* 115: 102–141.

Russell, C. I., Bar, K. A., and Adamson, L. B. (1997) Social referencing by young chimpanzees *(Pan troglodytes)*. *Journal of Comparative Psychology* 111: 185–193.

Ryle, G. (1949) *The Concept of Mind*. London: Hutchinson.

Sabbagh, M. A., and Callanan, M. A. (1998) Metarepresentation in action: 3, 4, and 5 year olds' developing theories of mind in parent-child conversations. *Developmental Psychology* 3: 491–502.

Sabbagh, M. A., and Taylor, M. (2000) Neural Correlates of "theory of mind" reasoning: An event-related potential study. *Psychological Science* 11: 46–50.

Sabini, J. (1992) *Social Psychology*. New York: W. W. Norton.

Sabini, J., and Schulkin, J. (1994) Biological realism and social constructivism. *Journal for the Theory of Social Behavior* 24: 207–217.

Sabini, J., and Silver, M. (1982) *Moralities of Everyday Life*. New York: Oxford University Press.

Sabini, J., and Silver, M. (1998) *Emotion, Characters, and Responsibility*. Oxford: Oxford University Press.

Saffran, E. M., and Schwartz, M. F. (1994) Of cabbages and things: Semantic memory from a neuropsychological perspective. A tutorial review. In *Attention and Performance XV,* edited by C. Umilt... and M. Moscovitch. Cambridge, Mass.: MIT Press.

Santangelo, S. L., and Folstein, S. E. (1999) Autism: A genetic perspective. In *Neurodevelopmental Disorders,* edited by H. Tager-Flusberg. Cambridge, Mass.: MIT Press.

Saper, C. B. (1982) Reciprocal parabrachial-cortical connections in the rat. *Brain Research* 242: 33–40.

Saper, C. B. (1996) Role of the cerebral cortex and striatum in emotional motor response. *Progress in Brain Research* 107: 537–550.

Sartre, J. P. (1948) *The Emotions.* New York: Philosophical Library.

Savage-Rumbaugh, E. S. (1986) *Ape Language: From Conditioned Response to Symbol.* New York: Columbia University Press.

Savage-Rumbaugh, E. S., and Lewin, R. (1994) *Kanzi: The Ape at the Brink of the Human Mind.* New York: Wiley.

Savage-Rumbaugh, E. S., Rumbaugh, D. M., Smith, S. T., and Lawson, J. (1980) Reference: The linguistic essential. *Science* 210: 922–925.

Savage-Rumbaugh, S., Shanker, S. G., and Taylor, T. J. (1998) *Apes, Language, and the Human Mind.* Oxford: Oxford University Press.

Saver, J. L., and Damasio, A. R. (1991) Preserved access and processing of social knowledge in a patient with acquired sociopathy due to ventromedial frontal damage. *Neuropsychologia* 29: 1241–1249.

Sawaguchi, T., and Kudo, H. (1990) Neocortical development and social structure in primates. *Primates* 31: 283–290.

Scaife, R., and Bruner, J. S. (1975) The capacity for joint visual attention in the infant. *Nature* 253: 265–266.

Schacter, D. L. (1992) Understanding implicit memory: A cognitive neuroscience approach. *American Psychiatrist* 47: 559–569.

Schacter, D. L. (1996) *Searching for Memory.* New York: Basic.

Schacter, D. L., and Cooper, L. A. (1993) Implicit and explicit memory for novel visual objects: Structure and function. *Journal of Experimental Psychology* 19: 995–1009.

Schaller, G. B. (1963) *The Mountain Gorilla: Ecology and Behavior.* Chicago: University of Chicago Press.

Schiffer, S. (1972) *Meaning.* Oxford: Clarendon.

Schiffer, S. (1987) *Remnants of Meaning.* Cambridge, Mass.: MIT Press.

Schneirla, T. C. (1966) Behavioral development and comparative psychology. *Quarterly Review of Biology* 41: 283–303.

Schreiner, L., and Kling, A. (1953) Behavioral changes following rhinencenphalic injury in the cat. *Journal of Neurophysiology* 16: 643–649.

Schuler, A., and Prizani, B. M. (1985) Echolalia. In *Communication Problems in Autism,* edited by E. Schopler and G. B. Mesibov. New York: Plenum.

Schulkin, J. (1992) *The Pursuit of Inquiry.* Albany: State University of New York Press.

Schulkin, J. (1996a) *The Delicate Balance.* Lanham, Md.: University Press of America.

Schulkin, J. (1996b) Pragmatism and the cognitive and neural sciences. *Psychological Reports* 78: 499–506.
Schulkin, J. (1999) *The Neuroendocrine Regulation of Behavior.* Cambridge: Cambridge University Press.
Schulkin, J., McEwen, B. S., and Gold, P. W. (1994) Allostasis, amygdala and anticipatory angst. *Neuroscience and biobehavioral reviews* 18: 385–396.
Schult, C. A., and Wellman, H. M. (1997)) Explaining human movements and actions: Children's understanding of the limits of psychological explanation. *Cognition* 62: 291–324.
Shultz, T. R., Wells, D., and Sarda, M. (1980) Development of the ability to distinguish intended actions from mistakes, reflexes, and passive movements. *British Journal of Social and Clinical Psychology* 19: 301–310.
Schutz, A. (1967) *The Phenomenology of the Social World.* Translated by G. Walsh and F. Lehnert. Evanston, Ill.: Northwestern University Press.
Schutz, A., and Luckmann, T. (1973) *The Structures of the Life-World.* Translated by R. M. Zaner and H. T. Engelhardt Jr. Evanston, Ill.: Northwestern University Press.
Schwaber, J. S., Kapp, B. S., Higgins, C. A., and Rapp, P. R. (1982) Amygdaloid and basal forebrain direct connections with the nucleus of the solitary tract and the dorsal motor nucleus. *Journal of Neuroscience* 2: 1424–1436.
Schwartz, S. P. (1977) *Naming, Necessity, and Natural Kinds.* Ithaca, N.Y.: Cornell University Press.
Schyns, P. G., Goldstone, R. L., and Thibaut, J. P. (1998) The development of features in object concepts. *Behavioral and Brain Sciences* 21: 1–54.
Scotus, D. ([1962] 1975) *Philosophical Writings.* Translated A. Wolter. Indianapolis: Bobbs-Merrill.
Searle, J. R. (1983) *Intentionality.* Cambridge: Cambridge University Press.
Searle, J. R. (1992) *The Rediscovery of the Mind.* Cambridge, Mass.: MIT Press.
Searle, J. R. (1998) *Mind, Language, and Society.* New York: Basic.
Sebeok, T. A., and Umiker-Sebeok, J. (1980) *Speaking of Apes: A Critical Anthology of Two-Way Communication with Man.* New York: Plenum Press.
Seligman, M. E. P. (1970) On generality of the laws of learning. *Psychological Review* 77: 406–418.
Sellars, W. (1963) *Science, Perception, and Reality.* Atlantic Heights, N.J.: Humanities.
Sellars, W. (1968) *Science and Metaphysics.* London: Routledge and Kegan Paul.
Sellars, W. (1997) *Empiricism and the Philosophy of Mind.* Cambridge, Mass.: Harvard University Press.
Seyfarth, R. M., Cheney, D. L., and Marler, P. (1980) Monkey responses to three different alarm calls: Evidence for predatory classification and semantic communication. *Science* 210: 801–803.
Shantz, C. U. (1975) The development of social cognition. In *Review of Child Development Research,* vol. 5, edited by E. M. Hetherinton. Chicago: University of Chicago Press.

Shapiro, L. A. (1997) The nature of nature: Rethinking naturalistic theories of intentionality. *Philosophical Psychology* 10: 309–322.

Shatz, M. (1994) *Children's Early Understanding of Mind: Origins and Development.* Edited By C. Lewis and P. Mitchell. Hillsdale, N.J.: Lawrence Erlbaum.

Shelton, J. R., and Caramazza, A. (1999) Deficits in lexical and semantic processing: Implications for models of normal language. *Psychonomic Bulletin and Review* 6: 5–27.

Shepard, R. N. (1992) The perceptual organization of colors: an adaptation to regularities of the terrestorial world. In *The Adaptive Mind,* edited by J. H. Barkow, L. Cosmides, and J. Tooby. Oxford: Oxford University Press.

Shepard, R. N., and Cooper, L. A. (1982) *Mental Images and Their Transformations.* Cambridge, Mass.: MIT Press.

Shepard, R. N., and Metzler, J. (1971) Mental rotation of three-dimensional objects. *Science* 171: 701–803.

Shettleworth, S. J. (1971) Constraints on learning. In *Advances in the Study of Behavior,* vol. 4, edited by D. S. Lehrman, R. A., Hinde, and E. Shaw. New York: Academic.

Shettleworth, S. J. (1999) *Cognition, Evolution, and Behavior.* Oxford: Oxford University Press.

Shultz, T. R., and Shamash, F. (1981) The child's conception of intending act and consequence. *Canadian Journal of Behavioural Science* 13: 368–382.

Shweder, R. A. (1991) *Thinking Through Cultures: Expeditions in Cultural Psychology.* Cambridge, Mass.: Harvard University Press.

Siegal, M., Carrington, J., and Radel, M. (1996) Theory of mind and pragmatic understanding following right hemispheric damage. *Brain and Language* 53: 40–50.

Siegel, B. (1996) *The World of the Autistic Child.* Oxford: Oxford University Press.

Sigman, M., and Ungerer, J. (1984) Attachment behaviors in autistic children. *Journal of Autism and Developmental Disorders* 14: 231–44.

Sigman, M., Mundy, P., Sherman, T., and Ungerer, J. (1986) Social interactions of autistic, mentally retarded, and normal children and their caregivers. *Journal of Child Psychology and Psychiatry* 27: 647–656.

Simon, H. A. (1974) *The Sciences of the Artificial.* Cambridge, Mass.: MIT Press.

Simon, H. A. (1982) *Models of Bounded Rationality.* Cambridge, Mass.: MIT Press.

Simpson, G. G. ([1949] 1964) *The Meaning of Evolution.* Reprint, New Haven: Yale University Press.

Skinner, B. F. (1938) *The Behavior of Organisms.* New York: Appleton-Century-Crofts.

Skinner, B. F. (1957) *Verbal Behavior.* New York: Appleton.

Sleeper, W. (1986) *The Necessity of Pragmatism.* New Haven: Yale University Press.

Small, S. L., Hart, J., Nguyen, T., and Gordon, B. (1995) Distributed representations of semantic knowledge in the brain. *Brain* 118: 441–453.

Smith, E. E., and Jonides, J. (1999) Storage and executive processes in the frontal lobes. *Science* 283: 1657–1661.

Smith, E. E., and Medin, D. L. (1981) *Categories and Concepts.* Cambridge, Mass.: Harvard University Press.

Smith, I. M., and Bryson, S. E. (1994) Imitation and action in autism: A critical review. *Psychological Bulletin* 116: 259–283.

Smith, I. M., and Bryson, S. E. (1998) Gesture imitation in autism. I. nonsymbolic postures and sequences. *Cognitive Neuropsychology* 15: 747–870.

Smith, J. E. (1970) *Themes in American Philosophy.* New York: Harper Torchbooks.

Smith, J. E. (1978) *Purpose and Thought: The Meaning of Pragmatism.* New Haven, Conn.: Yale University Press.

Smith, W. J. (1977) *The Behavior of Communicating.* Cambridge, Mass.: Harvard University Press.

Sodian, B., Taylor, C., Harris, P. L., and Perner, J. (1991) Early deception and the child's theory of mind. *Child Development* 62: 468–483.

Spelke, E. S., Breinliger, K., Macomber, J., and Jacobson, K. (1992) Origins of knowledge. *Psychological Review* 99: 605–632.

Spelke, E. S., Vishton, P., and Von Hofsten, C. (1994) Object perception, object-directed action, and physical knowledge in infancy. In *The Cognitive Neurosciences,* edited by M. S. Gazzaniga. Cambridge, Mass.: MIT Press.

Sperber, D., and Wilson, D. (1986) *Relevance: Communication and Cognition.* Oxford: Blackwell

Sperling, G. (1960) *The Information Available in Brief Visual Presentations.* Psychological Monographs 74, no. 11 (whole no. 498)

Sperry, R. W. (1961) Cerebral organization and behavior. *Science* 133: 1749–1757.

Spinoza, B. ([1668] 1955) *On the Improvement of the Understanding: The Ethics Correspondence.* Translated by R. H. M. Elwes. New York: Dover.

Spitzer, M. (1999) *The Mind within the Net.* Cambridge, Mass.: MIT Press.

Squire, L. R. (1987) *Memory and Brain.* New York: Oxford University Press.

Squire, L. R. (1998) Memory systems. *Chicago Academy of Science* 321: 153–156.

Squire, L. R., and Zola, S. M. (1996) Structure and function of declarative and nondeclarative memory systems. *National Academy of Sciences* 93: 13515–13522.

Stellar, E. (1954) The physiology of motivation. *Psychological Review* 61: 5–21.

Stellar, J., and Stellar E. (1985) *The Neurobiology of Reward.* New York: Springer.

Stephan, K. M., Fink, G. R., and Passingham, R. E. (1995) Functional anatomy of the mental representation of upper extremity movements in healthy subjects. *Journal of Neurophysiology* 73: 921–924.

Sternberg, S. (1969) Mental processes revealed by reaction time experiments. *American Scientist* 27: 815–819.

Stich, S. P. (1983) *From Folk Psychology to Cognitive Science: The Case Against Belief.* Cambridge, Mass.: MIT Press.

Stone, V. E., Baron-Cohen, S., and Knight, R. L. (1998) Frontal lobe contributions to theory of mind. *Journal of Cognitive Neuroscience* 105: 640–656.

Swanson, L. W., and Mogenson, G. J. (1981) Neural mechanisms for the functional coupling of autonomic, endocrine, and somatmotor responses in adaptive behavior. *Brain Research Reviews* 3: 1–34.

Swanson, L. W., and Petrovich, G. D. (1998) What is the amygdala? *Trends in Neuroscience* 21: 323–331.

Swerdlow, N. R., and Young, A. B. (1999) Neuropathology of Tourette Syndrome. *CNS Spectrums* 4: 65–81.

Tager-Flusberg, H. (1981) On the nature of linguistic functioning in early infantile autism. *Journal of Autism and Developmental Disorders* 11: 45–56.

Tager-Flusberg, H. (1985) The conceptual basis for referential word meaning in children with autism. *Child Development* 56: 1167–1188.

Tager-Flusberg, H. ([1993] 1999) What language reveals about the understanding of minds in children with autism. In *Understanding Other Minds,* edited by S. Baron-Cohen, H. Tager-Flusberg, and D. J. Cohen. Reprint, Oxford: Oxford University Press.

Tager-Flusberg, H., ed. (1999) *Neurodevelopmental Disorders.* Cambridge, Mass.: MIT Press.

Tager-Flusberg, H., Baron-Cohen, S., and Cohen, D. ([1993] 1999) An introduction to the debate. In *Understanding Other Minds,* edited by S. Baron-Cohen, H. Tager-Flusberg, and D. J. Cohen. Reprint, Oxford: Oxford University Press.

Tarski, A. (1944) The semantic conception of truth and the foundations of Semantics. *Philosophy and Phenomenological Research* 4 (1983). Der Wahrheitsbegriff in den formalisierten Sprachen (1925). *Studia Philosophica* 1; reprinted as The concept of truth in formalised languages, translated by J. H. Woodger, in *Logic, Semantics, Metamathematics: Papers from 1923–1938,* 2d. ed. edited by J. Corcoran. Indianapolis: Hackett.

Taylor, M. (1996) A theory of mind perspective in social cognitive development. In *Handbook of Perception and Cognition,* vol. 13, *Perceptual and Cognitive Development,* edited by R. Gelman, T. Au, E. C. Carterette, and M. P. Friedman. New York: Academic.

Taylor, M., and Carlson, S. M. (1997) The relation between individual differences in fantasy and theory of mind. *Child Development* 68: 436–455.

Teitelbaum, P., and Pellis, S. M. (1992) Toward a synthetic physiological psychology. *Psychological Science* 3: 4–20.

Teitelbaum, P., Teitelbaum, O., Nye, J., Fryman, J., and Mauer, R. G. (1998) Movement analysis in infancy may be useful for early diagnosis of autism. *Proceedings of the National Academy of Sciences* 23: 13982–13988.

Terrace, H., et al. (1979) Can an ape create a sentence? *Science* 206: 892–902.

Tinbergen, N. (1951) *The Study of Instinct.* Oxford: Clarendon.

Tinbergen, N. (1974) Ethology and stress diseases. *Science* 185: 20–27.

Tirassa, M. (1999) Communicative competence and the architecture of the mind/brain. *Brain and Language* 68: 419–441.

Tolman, E. C. (1932) *Purposive Behavior in Animals and Men.* New York: Appleton-Century-Crofts.

Tolman, E. C. (1948) Cognitive maps in rats and men. *Psychological Review* 55: 189–208.

Tomasello, M., and Akhtar, N. (1995) Two-year-olds use pragmatic cues to differentiate reference to objects and actions. *Cognitive Development* 10: 201–224.

Tomasello M., and Call, J. (1997) *Primate Cognition.* New York: Oxford University Press.

Tomasello, M., Call, J., Nagell, K., Olguin, R., and Carpenter, M. (1994) The learning and use of gestural signals by young chimpanzees: A trans-generational study. *Primates* 35: 137–154.

Tomasello, M., and Camaioni, L. (1997) A comparison of the gestural communication of apes and human infants. *Human Development* 40: 7–24.

Tomasello, M., Gust, D. A., and Evans, A. (1990) Peer interaction in infant chimpanzees. *Folia Primatologica* 55: 33–40,

Tomasello, M., Kruger, A. C., and Ratner, H. H. (1993) Cultural learning. *Behavioral and Brain Sciences* 16: 495–552.

Tomasello, M., Savage-Rumbaugh, E. S., and Kruger, A. C. (1993) Imitative learning of actions on objects by children, chimpanzees, and enculturated chimpanzees. *Child Development* 64: 1688–1705.

Tooby, J., and Cosmides, L. (1991) The psychological foundation of culture. In *The Adapted Mind: Evolutionary Psychology and the Generation of Culture,* edited by J. H. Barkow, L. Cosmides and J. Tooby. New York: Oxford University Press.

Tooby, J., and Cosmides, L. (1994) Mapping the evolved functional organization of mind and brain. In *The Cognitive Neurosciences,* edited by M. S. Gazzaniga. Cambridge: Bradford, MIT Press.

Townsend, J., Harris, N. S., and Courchesne, E. (1996) Visual attention abnormalities in autism: Delayed orienting to location. *Journal of International Neuropsychology* 6: 541–550.

Tranel, D., Damasio, H., and Damasio, A. R. (1997) A neural basis for the retrieval of conceptual knowledge. *Neuropsychologia* 35: 1319–1327.

Tranel, D., Logan, C. G., Frank,, R. J., and Damasio, A. R. (1997) Explaining category-related effects in the retrieval of conceptual and lexical knowledge for concrete entities: Operationalization and analysis of factors. *Neuropsychologia* 35: 1329–1339.

Treissman, A. M. (1964) Selective attention in man. *British Medical Journal* 20: 12–16.

Trivers, R. (1985) *Social Evolution.* Menlo Park, Calif.: Benjamin Cummings.

Tulving, E. (1985) How many memory systems are there? *American Psychologist* 40: 385–398.

Tulving, E., and Markowitsch, H. J. (1998) Episodic and declarative memory: Role of the hippocampus. *Hippocampus* 8: 198–204.

Tversky, B., and Hemenway, K. (1984) Object, parts, and categories. *Journal Experimental Psychology, General* 113: 169–193.

Ullman, S. (1980) Against direct perception. *Behavioral and Brain Sciences* 3: 375–385.

Ullman, M. T., Corkin, S., Coppola, M., Hickok, G., Growdon,. H., Horoshe, W. J., and Pinker, S. (1997) A neural dissociation within language: Evidence that the mental dictionary is part of declarative memory, and that grammatical rules are processed by the procedural system. *Journal of Cognitive Neuroscience* 9: 266–286.

Ungerleider, L. G., and Haxby, J. V. (1994). "What" and "where" in the human brain. *Current Opinion in Neurobiology* 4: 157–165.

Ungerleider, L. G., and Mishkin, M. (1982). Two cortical visual systems. In *Analysis of Visual Behavior,* edited by D. J. Ingle, M. A. Goodale, and R. J. W. Mansfield. Cambridge, Mass.: MIT Press.

Van Essen, D. C., Maunsell, J. H. R., and Bixby, J. L. (1981). The middle temporal visual area in the macaque: Myeloarchitecture, connections, functional properties, and topographic organization. *Journal of Comparative Neurology* 5: 165–204.

Van Lawick-Goodall, J. (1968a) The behaviour of free-living chimpanzees in the Gombe Stream reserve. *Animal Behaviour Monographs* 1: 161–311.

Varela, F. J., Thompson, E., and Rosch, E. (1991) *The Embodied Mind.* Cognitive Science and Human Experience series. Cambridge, Mass.: MIT Press.

Visalberghi, E., and Fragaszy, D. (1995) The behaviour of capuchin monkeys, *Cebus apella,* with novel food: The role of social context. *Animal Behaviour* 49: 1089–1095.

Vogt, S. (1996) Imagery and perception-action mediation in imitative actions. *Cognitive Brain Research* 3: 79–86.

Volkmar, F. R. (1998) *Autism and Pervasive Developmental Disorders.* Cambridge: Cambridge University Press.

Volkmar, F. R., Sparrow, S. S., Goudereau, D., Cicchetti, D. V., Paul, R., and Cohen, D. J. (1987) Social deficits in autism: An operational approach using the Vineland Adaptive Behavior Scales. *Journal of the American Academy of Child Psychiatry* 26: 156–161.

Von Vonin, G. (1960) *Some Papers on the Cerebral Cortex.* Springfield, Ill. Chalres C. Thomas.

Von Uexkull, J. (1934) A stroll through the world of an animal. In *Instinctive Behavior,* edited by K. Lashley. Madison, Ct.: International Universities Press.

Vygotsky, L. (1962) *Thought and Language.* Cambridge, Mass.: MIT Press. First published in Russian, 1926.

Vygotsky, L. (1978) *Mind in Society: The Development of Higher Psychological Processes.* Edited by M. Cole, V. John Steiner, S. Scribner, and E. Souberman. Cambridge, Mass.: Harvard University Press.

Vygotsky, L., and Luria, A. (1993) *Studies on the History of Behavior.* Hillsdale, N.J.: Lawrence Erlbaum.

Walden, T. A., and Ogan, T. A. (1988) The development of social referencing. *Child Development* 59: 1230–1240.

Warrington, E., and McCarthy, R. (1987) Categories of knowledge: Further fractionation and an attempted integration. *Brain* 110: 1273–1296.

Warrington, E., and Shalice, T. (1984) Category-specific semantic impairments. *Brain* 107: 829–854.

Waterhouse, L., and Fein, D. (1997) Perspectives on social impairment. In *Autism and Developmental Disorders: A Handbook,* edited by D. J. Cohen and F. R. Volknar. New York: Wiley.

Waterhouse, L., Fein, D., and Modahl, C. (1996) Neurofunctional mechanisms in autism. *Psychological Review* 103: 457–489.

Watson, J. B. (1914) *Behavior: An Introduction to Comparative Psychology.* New York: Holt, Rinehart and Winston.

Watson, J. B. (1925) *Behaviorism.* New York: W. W. Norton.

Weiskrantz, L. (1956) Behavioral changes associated with ablations of the amygdaloid complex in monkeys. *Journal of Comparative and Physiological Psychology* 49: 381–391.

Weiskrantz, L. (1997) *Consciousness Lost and Found: A Neuropsychological Exploration.* New York: Oxford University Press.

Weiss, G., and Haber, H. F. (1999) *Perspectives on Embodiment.* London: Routledge.

Wellman, H. (1990) *The Child's Theory of Mind.* Cambridge, Mass.: MIT Press.

Wellman, H. M., and Hickling, A. K. (1994) The mind's "I": Children's conception of the mind as an active agent. *Child Development* 65 (6): 1564–1580.

Wellman, H. G. (1998) Culture variation and levels of analysis in folk psychologies. *Psychological Bulletin* 123: 33–36.

Wellman, H. M., and Woolley, J. D. (1990) From simple desires to ordinary beliefs: The early development of everyday psychology. *Cognition* 35: 245–275.

Werner, H. (1948) *Comparative Psychology of Mental Development.* Chicago: Follett.

Westergaard, G. C., and Hyatt, C. W. (1994) The responses of bonobos *(Pan paniscus)* to their mirror images: Evidence of self-recognition. *Human Evolution* 9: 273–279.

Wetherby, A. M., and Prutting, C. A. (1984) Profiles of communicative and cognitive-social abilities in autistic children. *Journal of Speech and Hearing Research* 27: 365–387.

Whalen, P. J., Rauch, S. L., Etcoff, N. L., McInerney, S. G., Lee, M. B., and Jenike, M. A. (1997) Masked presentations of emotional facial expressions modulate amygdala activity without explicit knowledge. *Journal of Neuroscience* 18: 411–418.

Whitehead, A. N. (1926) *Science and the Modern World.* Cambridge: Cambridge University Press.

Whitehead, A. N. ([1927] 1953) *Symbolism*. Reprint, New York: Macmillan.
Whiten, A., ed. (1991) *Natural Theories of Mind*. New York: Oxford University Press.
Whiten, A. (1993) Evolving a theory of mind: The nature of non-verbal mentalism in other primates. In *Understanding Other Minds: Perspectives from Autism*, edited by S. Baron-Cohen, H. Tager-Flusberg, and D. J. Cohen. New York: Oxford University Press.
Whiten, A., and Byrne, R. W. (1988) Tactical deception in primates. *Behavioral and Brain Sciences* 11: 233–244.
Whiten, A., Goodal, J., McGrew, W. C., Nishida, T., Reynolds, V., Sugiyama, Y., Tutin, C. E. G., Wronghan, R. W., and Boesch, C. (1999) Cultures in chimpanzees. *Nature* 399: 682–685.
Williams, M. (1977) *Groundless Beliefs*. New Haven, Conn.: Yale University Press.
Wimmer, H., and Weichbold, V. (1994) Children's theory of mind: Fodor's heuristics examined. *Cognition* 53: 45–57.
Wing, L. (1981) Asperger syndrome: A clinical account. *Journal of Psychological Medicine* 11: 115–129.
Wing, L., ed. (1988) *Aspects of Autism: Biological Research*. London; Gaskell: Royal College of Psychiatrists.
Wing, L. ([1991] 1997) The relationship between Asperger syndrome and autism. In *Autism and Asperger Syndrome*, edited by U Frith. Reprint, Cambridge: Cambridge University Press.
Wing, L., and Gould, J. (1979) Severe impairments of social interaction and associated abnormalities in children: Epidemiology and classification. *Journal of Autism and Developmental Disorders* 9: 11–30.
Winner, E., Bronell, H., Happe, F., Blum, A., and Pincus, D. (1998) Distinguishing likes from jokes: Theory of mind deficits and discourse interpretation in right hemispheric brain-damaged patients. *Brain and Language* 62: 89–106.
Wise, R., Chollet, F., Hadar, U., et al. (1991) Distribution of cortical networks involved in word comprehension and word retrieval. *Brain* 114: 1803–1817.
Wittgenstein, L. ([1953] 1958) *Philosophical Investigations*. 2d ed. Translated by G. E. M. Anscombe. Oxford: Basil Blackwell.
Woodruff, G., and Premack, D. (1979) Intentional communication in the chimpanzee: The development of deception. *Cognition* 7: 333–362.
Woolley, J. D., and Wellman, H. M. (1990) Young children's understanding of realities, nonrealities, and appearances. *Child Development* 61: 946–961.
Woolley, J. D., and Wellman, H. M. (1993) Origin and truth: Young children's understanding of imaginary mental representations. *Child Development* 64: 1–17.
Wright, C. I., Peterson, B. S., and Rauch, S. L. (1999) Neuroimaging studies in Tourette syndrome. *CNS Spectrum* 4: 54–61.
Yerkes, R. M. (1916) *The Mental Life of Monkeys and Apes*. Delmar, N.Y.: Scholars' Facsimiles and Reprints.
Yerkes, R. M. (1927) *The Mind of a Gorilla*. Genetic Psychology Monographs 2: 1–193.

Yerkes, R. M. (1943) *Chimpanzees: A Laboratory Colony.* New Haven, Conn.: Yale University Press.

Young, A. W. (1998) *Face and Mind.* Oxford: Oxford University Press.

Young, L. J., Nilsen, R., Waymire, K. G., MacGregor, G. R., and Insel, T. R. (1999) Increased affiliative response to vasopressin in mice expressing the receptor from a monogamous vole. *Nature* 400: 766–768.

Young, L. J., Winslow, J. T., Wang, Z., Gingrich, B., Guo, Q., Matzuk, M. M., and Insel, T. R. (1997) Gene targeting approaches to neuroendocrinology: Oxytocin, maternal behavior, and affiliation. *Hormones and Behavior* 31: 221–231.

Zajonc, R. B. (1980) Feeling and thinking: Preferences need no inferences. *American Psychologist* 35: 151–175.

Zeki, S., Watson, J. D., and Frackowiak, R. S. (1993) Going beyond the information given: The relation of illusory visual motion to brain activity. *Proceedings of the Royal Society of London—Series B, Biological Sciences* 252: 215–222.

Zelazo, P. D., and Frye, D. (1998) Cognitive complexity and control: 11. The development of executive function in childhood. *Current Directions in Psychological Science* 7: 121–125.

Zilbovicius, M., Garreau, B., Samson, Y., Remy, P., Barthelemy, C., Syrota, A., and Lelord, G. (1995) Delayed maturation of the frontal cortex in childhood autism. *American Journal of Psychiatry* 152: 248–252.

Name Index

Adolphs, R., xvii, 1, 102, 114, 118, 120, 125, 129, 130
Aggleton, J. P., 118
Alheid, G. F., 117, 122
Allen, C., 38
Alpert, N. M., 8, 9
Amaral, D. G., 118
Anderson, R. A., 26
Annett, J., 111
Anscombe, G. E. M., 3
Arbib, M. A., xviii, 8, 12, 14, 16, 129
Aristotle, xi
Aronson, E., xv
Asperger, H., 88
Astington, J. W., xvi, 66, 75–77
Aulkainen, S., 129
Austin, J. L., 3, 6
Axelrood, J., 142n4

Bachevalier, J., 123
Bamshad, M., 118
Barkow, J. H., xii
Baron, J., 23, 104, 109
Baron-Cohen, S., xi, xv–xviii, 1, 19, 46, 54, 87, 88, 91–93, 95–108, 118, 127, 133
Barresi, J., 7
Bartsch, K., 67, 69–72, 83
Batramino, C. A., 117
Bauman, M. L., 105
Baverel, G., 129
Beach, F. A., 25
Bechara, A., 129, 130
Behmann, M., 10
Behrend, D. A., 65
Bekoff, M., 38
Berridge, K. C., 27, 84, 114, 116, 122–124
Boden, M. A., 5
Bogdan, R. J., xi, xvi, 4, 112, 123, 137
Bogen, J. E., 25
Bohman, J. F., 134
Bookheimer, S. Y., 12
Botterill, G., 74
Bourdieu, P., xii, 28, 83
Bowlby, J., 114
Bradley, M. M., 116
Bratman, M., 7
Breitter, H. C., 120
Brentano, F., 2
Bronell, H., 102, 138
Brothers, L., xvi–xviii, 1, 54, 104, 113, 116, 118, 120, 133
Brown, H. I., 5
Brown, P., xvii
Brunner, J., 59, 66, 72–73
Bryson, S. E., 109
Bucy, P. C., 118
Byrne, R. W., xvi, 34, 51, 52

Cador, M., 114, 116, 120
Cadwaller, T. C., 141n1
Call, J., xvi, 45, 49, 50, 52, 53, 68, 142n6
Calvert, G. A., 10
Cannon, W. B., 25
Caramazza, A., 65
Carey, S., xii, 14, 17, 18, 24, 59, 60, 63, 64, 66, 137, 141n3
Carlson, S. M., 76

Carpenter, M. M., 66, 85
Carruthers, P., 82
Carter, C. S., 134
Cassidy, K. W., 84
Chao, L. L., 14
Charman, T., 101
Cheney, D. L., xvi, 37–45
Chomsky, N., xii, xiv, 4, 22, 137
Churchland, P. M., 7, 137, 141n2, 142n5
Churchland, P. S., 7, 137, 141n2, 142n5
Clark, A., xii, xv–xvii, 1, 5–6, 16, 17, 24, 59, 60, 66, 111, 112, 125
Cohen, D., xvi
Cohen, M. S., 10
Coley, J. D., 60
Cooper, L. A., 8, 17
Corona, R., 109
Cosmides, L., xi
Courchesne, E., 105
Cutting, A. L., 136

Damasio, A. R., xi, xvii, 16, 105, 112, 120, 124–125, 129
Damasio, H., 125
Darwin, C., xiv, 29, 55, 114, 115
Dasser, V., 65, 67
Davidson, R. J., 114, 116
Davis, D. R., 52
Dawkins, R., 142n4
De Sousa, R., 113
Decety, J., xvii, 10, 12–14, 111, 129
Dennett, D. C., xi, xiii, xv, 3, 5, 6, 24, 41, 136
DeOlmost, J., 117
Descartes, R., xii
Desimone, R., 54
DeWaal, F. B. M., 51, 56, 134, 135
Dewey, J., xiii–xiv, 7, 14, 19–21, 23, 83, 113, 123, 137, 141n1
Dewsbury, D. A., 26
Dickinson, A., 22–23, 31, 52, 84
Dolgin, K. G., 65
Domesick, V. B., 114, 117, 120
Dominey, P., 123
Dretske, F. I., 6
Duchenne, B., 114
Dufty, A., 141n4
Dunbar, R. I. M., 43
Durozard, D., 129

Eddy, T. J., xvi, 45–49, 52, 53
Ekman, P., 55, 114
Elster, J., 142n4
Emery, N. J., xvii, 12, 118, 133, 141n3
Ericsson, K. A., 24
Erneling, C. E., 27
Ervin, F. R., 61
Estes, D., 83
Evans, C., xv, 43
Everitt, B. J., 114, 116, 120

Farah, M. J., 8
Fein, D., 105, 108
Fein, D. A., 44
Fentress, J. C., 123
Fink, G. R., 10
Fiske, A. P., 6, 28
Flanagan, O., 134
Flavell, E. R., 71, 74, 75
Flavell, J. H., xvi, 60, 74, 75, 94
Fletcher, P. C., 102, 103, 105
Fodor, J. A., 23, 24, 27, 28, 60, 83
Folstein, S. E., 88
Fonberg, E., 117–118, 120
Fox, N., 125
Fox, N. A., 116
Frackowiak, R. S. J., 10
Frege, G., 2
Fridlund, A., xiv
Frijda, N., 113
Frith, C. D., xvii, 99, 102
Frith, U., xvi, xvii, 87–89, 91, 93, 95, 97, 99–105, 107
Frye, D., 66

Gadamer, H. G., 137
Galanter, E., 22
Gallagher, M., 104, 118

Gallese, V., 16, 111
Gallistel, C. R., xii, xv, xvii, 26, 113, 114
Gallup, G. G., Jr., 46, 52
Garcia, J., 17, 61
Gardner, B. T., 36
Gardner, H., 19, 36
Gazzaniga, M. S., 19, 25, 26, 125
Geertz, C., xiv, 137
Gelman, S. A., 59, 60, 63, 64
Geschwind, N., 26
Gibson, E., 1
Gibson, J. J., xii, 16–17, 60, 114
Glanzer, M., 23
Goel, V., 102, 127–129
Goffman, E., 6, 52, 137
Gold, P. W., 118
Goldman, A., 16, 82, 111
Goldman-Rakic, P. S., 125
Gomez, J. C., xvi, 50, 51
Goodall, J., 50
Gopnik, A., xvi, 59, 66, 72–77
Gray, T. S., 116
Graybiel, A. M., xvii, 59, 84, 122–125
Green, F. L., 74, 75
Grice, H. P., 6, 7
Griffin, D. R., 22, 57
Gross, C. G., 54
Gulyas, B., 10

Habermas, J., 4, 37
Hamilton, W. J., III, 20
Hankins, W. G., 17
Hanson, N. R., xiii, xiv, 5, 74
Happe, F., 102, 104, 105, 138
Hari, R., xvii, 10
Harnard, S., 112
Harris, N. S., 82, 105
Harris, P. L., xvi, 59, 82, 84, 91, 107, 108
Haugeland, J., 134
Hauser, M. D., xv, xvi, 14, 17, 43, 54, 56, 66, 137, 141n3
Haxby, J. V., 55, 120
Hayes, C. H., 36, 50
Hayes, K. J., 36, 50
Hebb, D. O., 22, 25
Heelan, P. A., xii, xv, xvii, 6, 17, 74, 111, 112
Heimer, L., 122
Hempel, C. G., 65
Herbert, J., 134
Herrick, C. J., 116, 117, 125
Heyes, C. M., xvi, 31, 46, 55, 84
Hickling, A. K., 80, 81
Hietanen, J. K., 55
Hiley, D. R., 134
Hinde, R. A., xi, 25, 55
Hirschfeld, L. A., 59
Hobson, P., 88
Holland, P. C., 118
Holt, D. J., 122
Hopkins, W. D., 50, 54
Hughes, C., 136
Humphrey, N. K., 43, 44, 137
Husserl, E., 2

Insel, T. R., 133, 134
Ishai, A., 8, 55

Jackson, J. H., 87
Jaeger, J. J., 124
James, W., xiii, 19–21, 23, 66, 111, 113
Jaspers, K., 1, 6
Jeannerod, M., xvii, 8, 10, 13, 14, 16, 83, 111, 129
Jenkins, J. M., 76
Jennings, H. J., 142n8
Joanisse, M. F., 124
Joas, H., 141n1
Johnson, D. M., 27
Johnson, M., xi, xvii, 1, 4, 7, 14, 28, 59, 60, 112, 123
Johnson, S. C., 63
Johnston, J. B., 116
Jolly, A., xi, 43, 44
Jonides, J., 125
Joseph, 84, 136
Juno, C., 51

Kagan, J., 59, 143n11
Kahneman, D., 23
Kanner, L., 88
Kant, I., xii, 113
Karmiloff-Smith, A., 60
Keil, F. C., 18, 59, 60, 62, 63
Kelley, A. E., 114, 120, 122, 123
Kemper, T. L., 105
Kitcher, P., xii
Kling, A., 118
Kluver, H., 22, 118
Knight, R. L., 127
Knowlton, B. J., 28, 124
Kohler, W., 22, 34
Konorski, J., 22,112, 116
Kornblith, H., xiv, 17, 60
Kosslyn, S. M., 8, 9, 26
Kostarczyk, E., 118
Krebs, J. R., 142n4
Kretak, J. E., 117
Kruger, A. C., 83
Kunda, Z., 19

Lakoff, G., xi, xvii, 4, 7, 14, 28, 59, 112, 123
Lane, R. D., xvii, 113
Lang, P. L., 116
Lanting, G., 134, 135
Lashley, K. S., xvii, 22, 25, 111, 112, 123
Leavens, D. A., 50, 54
Lederhendler, I., 26
LeDoux, J. E., 113, 117, 120, 142n9
Lee, K., 99
Leekam, S. R., 76, 78, 93, 98
Leslie, A. M., xvi, 66, 82, 87, 91–100, 107
Lewin, R., 36
Lewis, D., 74
Lieberman, M. D., 28
Lillard, A., 84, 136, 137
Lillard, A. S., 83
Loeb, J., 142n8
Lorenz, K., 25
Luria, A., 83
Luria, A. R., 125, 129
Lyons, W., 2, 4, 84, 143n10

Mackintosh, N. J., 22, 24
Magnuson, D., 116
Malamut, B., 123
Mangels, J. A., 124
Markowitsch, H. J., 8
Marler, P., xv, 20, 34, 36, 41, 43, 114, 141n4
Marr, D., 137
Marsden, C. D., xvii, 111, 122, 124
Martin, A., xvii, 8, 12, 14, 15, 111
Matthews, G. B., 74
Maurer, R. G., 105, 125
Mayr, E., 142n7
McCarthy, R. A., 8
McEwen, B. S., 117, 118
McFarland, D., 24
McGinty, J. F., 120
Mead, G. H., xiv, 1, 19, 21, 27, 28, 83, 137, 141n1
Meltzoff, A. N., xvi, 4, 53, 73, 74
Menzel, E. W., Jr., 51
Merleau-Ponty, M., xi, 14, 112
Metzler, J., 8
Miller, N. E., 22, 25, 112, 114
Miller, P. H., 75
Millikan, R. G., 6, 63
Milner, B., 25, 129
Milner, P. M., 25
Mishkin, M., 123
Mistlin, A., 55, 56
Modahl, C., 105, 134, 136
Mogenson, G. J., 26, 114, 120
Moore, C., 7, 66
Moore, M., 53
Morgan, C., 25
Morris, J. S., 119, 120
Morton, J., xvi
Moscovitch, M., 10
Mounce, H. O., 141n1

Mountcastle, V. B., 125
Mountz, J. M., 105

Nagel, T., 57
Nagell, K., 66, 85
Nauta, W. J. H., 114, 117, 120
Neisser, U., 23
Neville, R. C., xiii, 133, 141n1
Newell, A., 22
Nishimori, K., 134
Norgren, R., 116, 117

Ogden, D. L., 82
Olds, J., 25
Olson, D. R., xvi
Oram, M. W., 54

Panksepp, J., 112–114, 134
Papez, J. W., 116, 117
Parr, H., 123
Parrott, W. G., xvii, 24, 27, 28, 113, 142n9
Parsons, L. M., 10
Pasculavaca, D. M., 108
Pashler, H. E., 24
Passingham, R. E., 10
Pavlov, I. P., 21
Peirce, C. S., xii–xiv, 7, 16, 23, 27, 113, 137, 141n1
Pellis, S. M., 26
Pepperberg, I., 142n4
Perani, D., 10, 12, 63
Perner, J., xvi, 76, 78, 81, 82, 91, 93, 98
Perrett, D. I., xvii, 12, 54–57, 118, 133, 141n3
Peterson, B. S., 108
Petrides, M., 129
Petrovich, G. D., 117, 122
Pfaff, D. W., 112
Philips, W., 108
Piaget, J., 6, 24, 63, 64, 71
Pickert, R., 141n4
Pinker, S., xi–xiii, 4, 22, 137
Piven, J., 105
Plotkin, H., xiii
Posner, M. I., 24, 26
Povinelli, D. J., xvi, 45–49, 52, 53
Premack, A. J., 31, 32, 65, 66
Premack, D., xi, xvi, 14, 30–32, 34–37, 63, 65–67, 99, 141n3
Preuss, T. M., 24, 49
Pribram, K. H., 22
Price, J. L., 117
Putnam, H., 27
Pylyshyn, Z., 28, 60

Quiatt, D., 53
Quine, W. V. O., 22, 23, 28, 61, 141n1

Ratner, C., 83
Reber, P. G., 28
Rescorla, R. A., 22, 23, 52
Rey, G., 2, 22, 27, 113, 141n
Reynolds, V., 53
Richter, C. P., 22, 25
Rickman, M., 116
Ring, B., 118
Ring, H. A., 104
Ristau, C. A., 14, 17, 18, 46, 57, 61
Rizzolatti, G., xviii, 8, 12, 14, 16, 129
Robbin, T. W., 114, 116, 120
Robinson, D. N., 23
Roland, P. E., 10
Rolls, E. T., xvii, xviii, 8, 12, 26, 54, 55, 118, 122, 129, 133
Romanes, G. J., 29
Rorty, R., 111, 141n1
Rosen, J. B., 104, 120
Roth, D., 93, 94
Rozin, P., xi–xiii, 7, 17, 18, 28, 44, 56, 61, 142n7
Ruffman, T., 76, 78
Rumelhart, D. E., xii
Runciman, W. G., 43
Rusinak, K. W., 17
Russell, B., 2, 70
Rutter, M., 108
Ryle, G., 3, 123

Sabbagh, M. A., 102, 126
Sabini, J., xiv, xv, xvii, 19, 27, 108, 113, 114, 137
Santagelo, S. L., 88
Saper, C. B., 122
Sartre, J-P., 113
Savage-Rumbaugh, E. S., 36, 50
Schacter, D. L., 17
Schiffer, S., 7
Schmeck, H. M., 117
Schneirla, T. C., 114, 116
Schreiner, L., 118
Schulkin, J., xii, xiv, xv, xvii, 6, 17, 20, 24, 26–28, 30, 74, 104, 112, 113, 117, 118, 120, 133, 134, 137, 141n1, 142n9
Schult, C. A., 77, 79, 80
Schutz, A., 1, 28
Schwaber, J. S., 116
Searle, J. R., xi, 5–7
Sebeok, T. A., 36
Seidenberg, M. S., 124
Seligman, M. E. P., 18, 61
Sellars, W., 2, 3, 27, 28, 142n9
Seyfarth, R. M., xvi, 37–45
Shalice, T., 8
Shanker, S. G., 36
Shanks, D., 23
Shantz, C. U., 83
Shatz, M., 74, 83
Shelton, J. R., 65
Shepard, R. N., 5, 8
Shettleworth, S. J., xv, xvi, 18, 24, 57, 61, 112
Shusterman, R., 134
Shweder, R. A., xv, 28
Siegal, M., 108
Sigman, M., 88
Silver, M., xiv, 113, 137
Simon, H. A., xii, 22–24
Skinner, B. F., 22, 25
Slaughter, V., 72
Sleeper, W., 141n1
Small, S. L., 8
Smith, E. E., 125
Smith, I. M., 109
Smith, J. E., xiii, 37, 133, 141n1
Smith, W. J., xv, 114
Spelke, E. S., 59, 63
Sperling, G., 23
Sperry, R. W., 25
Spinoza, B., 113, 139
Spitzer, M., 8
Squire, L. R., 8, 24, 28, 123, 124
Stellar, E., 25, 112, 114, 116
Stellar, J., 114, 116
Stephan, K. M., 10
Sternberg, S., 23
Stich, S. P., 7, 142n5
Stone, V. E., 127
Sutton, S. K., 114
Swanson, L. W., 26, 114, 117, 120, 122
Swerdlow, N. R., 125

Tager-Flusberg, H., xvi, 91, 94
Taylor, M., 36, 76, 82, 102
Taylor, S. E., 6
Teitelbaum, P., 26, 109
Terrace, H., 36
Thompson, W. L., 8, 9
Tinbergen, N., 25, 88
Tirassa, M., 4
Tolman, E. C., 22, 29
Tomasello, M., xvi, 45, 49, 50, 52–54, 66–68, 83, 85, 142n6
Tooby, J., xi
Tovee, M. T., 8, 54, 55
Tranel, D., 12, 120, 125, 129
Treissman, A. M., 24
Treves, A., xviii, 8, 26, 54, 55, 118, 122, 129, 133
Trivers, R., 142n4
Tulving, E., 8
Tversky, A., 23

Ulbaek, I., 67
Ullman, M. T., 124, 126
Umiker-Sebeok, J., 36

Ungerer, J., 88
Ungerleider, L. G., 55

Varela, F. J., 112
Vishton, P., 63
Vogt, S., 111
Volkmar, F. R., 88
Von Hofsten, C., 63
Von Vonin, G., 125
Vygotsky, L., 83

Wagner, A. R., 22, 23, 52
Warrington, E., 8
Waterhouse, L., 104, 108
Watson, J. B., 25
Watson, J. D., 10
Weiskrantz, L., 118
Wellman, H. M., xvi, 67, 68, 70–72, 77, 79–81, 99, 137
White, N. M., 25
Whitehead, A. N., 66, 111
Whiten, A., 51–53, 142n6
Wimmer, H., 98
Wing, L., 101
Winner, E., 102, 107, 108, 138
Winocur, G., 10
Wise, R., 8
Wittgenstein, L., xiv, 1, 3, 5, 137
Woodruff, G., xi, 30, 35, 37
Woolley, J. D., 69, 71, 79
Wright, C. I., 125

Yansen, J., 20, 30
Young, A. B., 125
Young, A. W., 55
Young, L. J., 133, 134

Zajonc, R. B., 113, 142n9
Zeki, S., 10
Zilbovicius, M., 105
Zola, S. M., 8, 28, 124

Subject Index

Abduction, xiii, xiv, 34
Action, 111
basal ganglia and neuropsychology of, 120, 122–125
intentionality in organization of, xi
Alarm calls, 45
Alliance formation in monkeys, 38, 39
Alzheimer's disease patients, 126
Amygdala, 116–120
Animals. *See also* Primates; *specific species and specific topics*
interpretation of world, xv–xvi
Animate objects
and causation, 65–66
distinguishing between inanimate and, 63–65
Animism and animacy, 63–65
Appearance. *See also* Imagination
distinguishing reality from, 71–74, 76, 77
Approach and avoidance behavioral responses, 114, 116, 120, 121
Asperger syndrome, 101, 102, 104
Attention, shared/joint, 99–101
Autism, 87–88
characteristic traits, 88, 89
DSM-IV criteria, 88, 90
as impairment in social knowledge, 94–97, 108
and knowledge of others' experience, 91–94, 107–109
shared attention and, 99–101
Autistic children, 107–109
true and false beliefs in, 98–99
Autistic people, xvi, 107–109
brain scans of and brain activity in, 103–106

Basal ganglia, 120, 122–124
Begging gestures, 46–47
Behaviorism, 21–23, 31, 36, 49, 56–57
Behaviorist vs. mentalist frameworks, children's choice of, 75–79
Belief- and desire-based language, children's use of, 67, 69–70
Beliefs
desires and, 70–71, 77, 83, 126–127
false, 73–74, 76, 77, 85, 126
in autistic children, 93, 98–99
Bodily motion
brain, faces, and, 54–55
Body, 111–112

Causal efficacy, 66
Causation, animate vs. inanimate, 65–66
Children, 82–83. *See also specific topics*
Chimpanzee work, doubts about, 45–52
Chimpanzees, 30, 31, 52
as having ideas and insight, 34
Cognitive adaptation, xi–xiii
"Cognitive," defined, 23
Cognitive development, conceptions of, 59
Cognitive revolution, 19, 24, 27–28
Cognitive science, 21–25
Cognitivism, xiii–xiv

Communicative social competence, 4
 in vervet monkeys, 37–43
Context and knowledge, 81–82

Deception, 43, 51–52, 98–99
Decision making, 23
Deduction, xiii
Desire- and belief-based language, children's use of, 67, 69–70
Desires, 71
 and beliefs, 70–71, 77, 83, 126–127

Elderly persons, cognitive performance of, 138–139
"Embodied mind," 112
Emotions
 cognitive/functionalist view of, 113–114
 neuroscience of, 114, 116–120
 social cognition, amygdala, and, 116–120
 theories of, 112–114
Empathy task, 101. *See also* Experience, others'
Experience, 21, 24–25. *See also specific topics*
 others'
  autism and knowledge of, 91–94, 107–109
  capturing and understanding, xvi–xvii, 19–21, 26–27
Extension, 2–3
Eye contact and gaze, 46–49, 52

Faces. *See also* Identification
 neurobiological responses to, 54–55
False-belief tasks, 73–74, 76, 77, 93, 98–99, 126
Fear, 120
Frontal cortex, 125
 knowledge of others and the, 125–129
 visceral anticipatory responses and the, 129, 130
Functionalism, 27

Gaze. *See* Eye contact and gaze
Gorillas, 50–51

Identification (recognition), 12–15
 in monkeys, 38, 40
Imaginary play, 84, 91–92
Imagination, 71–72
 vs. knowledge, 79–81
Induction, xiii, xiv
Information-processing neural systems, xvii–xviii, 130. *See also specific topics*
Inquiry, xiii–xiv
Intension, 2, 3
Intention, 2
 and meaning, 6–7
Intentional action, xvii–xviii, 16
Intentional attribution(s), 1, 4–7, 83, 84, 125
 in chimpanzees, 32, 35–36, 49, 50
Intentional discourse, 3
Intentional model of cultural learning, 53, 54
Intentional stances, 5, 7
Intentional systems, 4–6, 41
Intentionality, 17
 in children and infants, 66–71, 78–79, 83
 historical roots of the concept of, 2–4
 intrinsic vs. extrinsic states of, 5
 nature of, 1–2
 neurobiology of, xvii, 5
 in organization of action, xi
 in primates, 31–32, 41, 49, 50
 and social parsing, 134–138
"Introspective mind," 137
"Introspective organ," 107

Knowledge. *See also* Social knowledge; *specific topics*
 context and, 81–82
 shared, xiv
 vs. imagination, 79–81

Language, 22, 36, 94–95
and intentionality, 3–4, 84
paradigmatic cognitive function of, 124
Learning
cognitive mechanisms in, 7–8
intentional model of cultural, 53, 54
Learning theory(ies), 22–23
Lying, 98–99

Meaning, 1–3
brain function and, 7–16
in monkeys, 41, 42
unloading it from mind/brain back to the world, 16–17
"Means-and-goal" task, 66
Memory, cognitive mechanisms in, 7–8
Mentalism, 84–85
Mentalist vs. behaviorist frameworks, children's choice of, 75–79
Metaphors, 81
Metarepresentations, 93
Mind, theories of, xvi, 53, 55, 57, 102
autistic individuals', 93, 101, 103–107
children's, 71–72, 76–77, 81, 93, 101
neurobiology and, 101–107
Mind-body split, 111
Monkeys
communicative social competence and meaning in, 37–43
vervet, 38–45
Motivation, 25–26
Motor imagery in left hemisphere, 10, 12, 13

Natural kinds/categories, 60–63
Neocortex, 125, 128
Neuroscience. *See also specific topics*
behavioral, 26
cognitive, 26, 28
"Noncognitive" functions, 49, 52

Orangutans, 50
Orbito-frontal region, 127. *See also* Frontal cortex
Oxytocin gene expression, 133, 134, 136

Parkinson's patients, 123, 124
Past-tense production task, 126
Perception, 5, 6
neurobiology of, 8–12
visual, 73
Physiological psychology, 25–26
Picture-story sequencing task, 95–97
Planning, 129
Play, imaginary/pretend, 84, 91–92
Pragmatists and pragmatism, xiii, xiv, 19, 21, 26, 141n1
Preadaptation, 142n7
Prefrontal cortex, 129
Primates, higher, 27
theory of mind in, 30–37
Problem solving, xii, 23, 34
Psychobiology, 26

Reality, distinguishing appearance from, 71–74, 76, 77
Reason, xii, 23. *See also* Social reason(ing)
Recognition. *See* Identification; Self-recognition
Reconciliation in monkeys, 40
Representational change, 76, 77
Representation(s), 2, 16, 54, 70, 72
alternative and conflicting, 75–76
of bodily events, 112
of situations, understanding, 81–82
Rivalry in monkeys, 41

Self-recognition, 46
Semantics, 8
Semiotics, 37
Situational attitudes, 81
Social cognition, 6, 44
amygdala, emotions, and, 116–120

Social-cognitive ability, and intentional model of cultural learning, 53, 54
Social constructivism, xiv
Social intelligence, and primate evolution, 43–45
Social intentional relationships, 6
Social knowledge, 7, 44, 76, 83, 85, 125, 129
 autism as impairment in, 94–97, 108
 neurobiology, 125–129, 133
Social parsing, 107
 intentionality and, 134–138
Social reason(ing), xiv–xvi, xviii, 1. *See also* Reason
Somatic marker hypothesis, 112
Starlings, 45
Stream of consciousness, 74–75

Temporal cortex, 56
Thinking. *See also specific topics*
 children's thoughts about, 74–75
Tourette's syndrome, 101, 102
Trustworthiness ratings, 120, 121

Unconscious, cognitive, 56
Unconscious mechanisms, 43, 46, 52, 55

Visual perception, development of the understanding of, 73